ARTISTIC ANATOMY OF ANIMALS

ARTISTIC ANATOMY OF ANIMALS

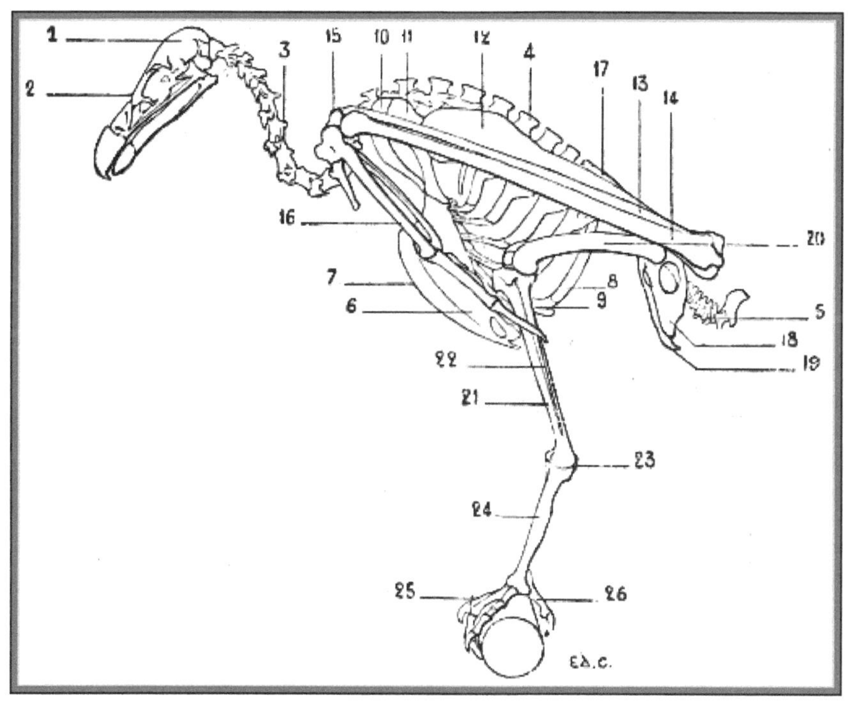

Édouard Cuyer

ILLUSTRATED &
PUBLISHED BY
E-KİTAP PROJESİ & CHEAPEST BOOKS

www.cheapestboooks.com

Istanbul

ISBN: 978-605-7876-71-3

TABLE OF CONTENTS

PREFACE ***7***

INTRODUCTION ***13***

CHAPTER I ***16***

OSTEOLOGY AND ARTHROLOGY **16**

THE TRUNK 16

THE POSTERIOR LIMBS[14] 105

THE POSTERIOR LIMBS IN SOME ANIMALS. 120

THE SKULL OF BIRDS 161

CHAPTER II ***163***

MYOLOGY **163**

THE MUSCLES OF THE TRUNK 165

MUSCLES OF THE ANTERIOR LIMBS 197

MUSCLES OF THE POSTERIOR LIMBS 239

MUSCLES OF THE HEAD 272

CHAPTER III ***288***

EPIDERMIC PRODUCTS OF THE TERMINAL EXTREMITIES OF THE FORE AND HIND LIMBS **288**

CHAPTER IV ***306***

PROPORTIONS **306**

PROPORTIONS OF THE HEAD OF THE HORSE[63] 320

CHAPTER V ***331***

THE PACES OF THE HORSE **331**

THE

ARTISTIC ANATOMY OF ANIMALS

PREFACE

A few lines will suffice to explain why we have compiled the present volume, to what wants it responds, and what its sphere of usefulness may possibly embrace.

In our teaching of plastic anatomy, especially at the École des Beaux-Arts—where, for the past nine years, we have had the very great honour of supplementing the teaching of our distinguished master, Mathias Duval, after having been prosector for his course of lectures since 1881—it is our practice to give, as a complement to the study of human anatomy, a certain number of lessons on the anatomy of those animals which artists might be called on to represent.

Now, we were given to understand that the subject treated in our lectures interested our hearers, so much so that we were not surprised to learn that a certain number repeatedly expressed a desire to see these lectures united in book form.

To us this idea was not new; for many years the work in question had been in course of preparation, and we had collected materials for it, with the object of filling up a void of which the existence was to be regretted. But our many engagements prevented us from executing our project as early as we would have wished. It is this work which we publish to-day.

Putting aside for a moment the wish expressed by our hearers, we feel ourselves in duty bound to inquire whether the utility of this publication is self-evident. Let it be clearly understood that we wish to express here our opinion on this subject, while putting aside every personal sentiment of an author.

No one now disputes the value of anatomical studies made in view of carrying out the artistic representation of man. Nevertheless—for we must provide against all contingencies—the conviction on this subject may be more or less absolute; and yet it must possess this character in an intense degree in order that these studies may be profitable, and permit the attainment of the goal which is proposed in undertaking them.

Fig. I.—Reproduction of a Sketch by Barye (Collections of the Anatomical Museum of the École des Beaux-Arts—Huguier Museum).

It is in this way that we ever strive to train the students whose studies we direct; not only to admit the value of these studies, but to be materially and deeply convinced of the fact without any restriction. Such is the sentiment which we endeavour to create and vigorously encourage. And we may be permitted to add that we have often been successful in this direction.

Therefore it is that, at the beginning of our lectures, and in anticipation of possible objections, we are accustomed to take up the question of the utility of plastic anatomy. And in so doing, it is in order to combat at the outset the idea—as mischievous as it is false—which is sometimes imprudently enunciated, that the possession of scientific knowledge is likely to tarnish the purity and freshness of the impressions received by the artist, and to place shackles on the emotional sincerity of their representation.

Fig. II.—Reproduction of a Sketch of Barye (Collections of the Anatomical Museum of the École des Beaux-Arts—Huguier Museum).

It is chiefly by employment of examples that we approach the subject. These strike the imagination of the student more forcibly, and the presentation of models of a certain choice, although rough in execution, is, in our opinion, preferable to considerations of an order possibly more exalted, but of a character less clearly practical. Let us, then, ask the question: Those artists whose eminence nobody would dare to question, did they study anatomy? If the answer be in the affirmative, we surely cannot permit ourselves to believe that we can dispense with a similar course. And, as proof of the studies of this class which the masters have made, we may cite Raphael, Michelangelo, and, above all, Leonardo da Vinci; and, of the moderns, Géricault. And we may more clearly define these proofs by an examination of the reproductions of their anatomical works, chosen from certain of their special writings.[1]

[1] Mathias Duval and A. Bical, 'L'anatomie des Maîtres.' Thirty plates reproduced from the originals of Leonardo da Vinci, Michelangelo,

Raphael, Géricault, etc., with letterpress and a history of plastic anatomy, Paris, 1890.

The manuscripts of Leonardo da Vinci of the Royal Library, Windsor, 'Anatomy, Foliæ A.,' published by Théodore Sabachnikoff, with a French translation, written and annotated by Giovanni Piumati, with an introduction by Mathias Duval. Édouard Rouveyre, publisher, Paris, 1898.

Mathias Duval and Édouard Cuyer, 'History of Plastic Anatomy: The Masters, their Books, and Anatomical Figures' (Library of Instruction of the School of Fine Arts), Paris, 1898.

Accordingly, there is no scope for serious discussion, and it only remains for us to enunciate the opinion that it is necessary that we should imitate those masters, and, with a sense of respectful discipline, follow their example.

Here, with regard to the anatomy of animals, we pursue the same method, and the example chosen shall be that of Barye. His talent is too far above all criticism to allow that this example should be refused. The admiration which the works of this great artist elicit is too wide-spread for us to remain uninfluenced by the lessons furnished by his studies. It is sufficient to see the sketches relating to these studies, and his admirable casts from nature which form part of the anatomical museum of the École des Beaux-Arts, to be convinced that the artistic temperament, of which Barye was one of the most brilliant examples, has nothing to lose by its association with researches the precision of which might seem likely to check its complete expansion.

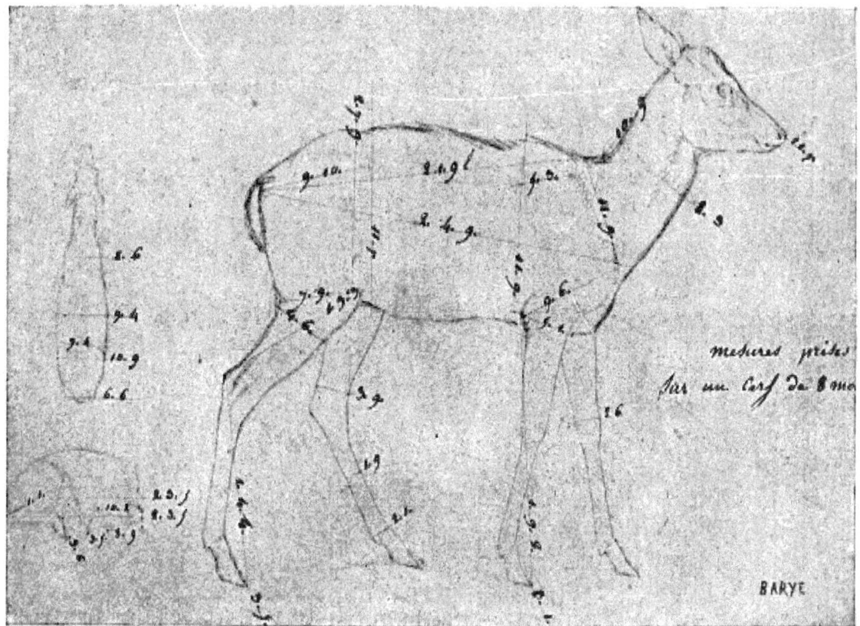

Fig. III.—Reproduction of a Sketch of Barye (Collections of the Anatomical Museum of the École des Beaux-Arts—Huguier Museum).

In those sketches we find proofs of observation so scrupulous that we cannot restrain our admiration for the man whose ardent imagination was voluntarily subjected to the toil of study so profound.

If the example of Barye, with whom we associate the names of other great modern painters of animals, can determine the conviction which we seek to produce, we shall be sincerely glad. To contribute to the propagation of useful ideas, and to see them accepted, gives a feeling of satisfaction far too legitimate for us to hesitate to say what we should feel if our hope be realized in this instance.

ÉDOUARD CUYER.

Fig. IV.—Reproduction of a Sketch of Barye

(Collections of Anatomical Museum of the School of Fine Arts—Huguier Museum).

INTRODUCTION

GENERALITIES OF COMPARATIVE ANATOMY

Of the animals by which we are surrounded, there are some which, occupying a place in our lives by reason of their natural endowments, are frequently represented in the works of artists—either as accompanying man in his work or in his amusements, or as intended to occupy the whole interest of the composition.

The necessity of knowing, from an artistic point of view, the structure of the human body makes clear the importance we attach, from the same point of view, to the study of the anatomy of animals—that is, the study of comparative anatomy. The name employed to designate this branch of anatomy shows that the object of this science is the study of the relative position and form which each region presents in all organized beings, taking for comparison the corresponding regions in man. The head in animals compared with the human head; the trunk and limbs compared to the trunk and limbs of the human being—this is the analysis we undertake, and the plan of the subject we are about to commence.

Our intention being, as we have just said, the comparison of the structure of animals with that of man, should we describe the anatomy of the human being in the pages which follow? We do not think so. Plastic human anatomy having been previously studied in special works,[2] we take it for granted that these have been studied before undertaking the subject of comparative anatomy. We will therefore not occupy time with the elementary facts relative to the skeleton and the superficial layer of muscles. We will not dilate on the division of the bones into long, short, large, single, paired, etc. All these preliminary elements we shall suppose to have been already studied.

This being granted, it is, nevertheless, necessary to take a rapid bird's-eye view of organized beings, and to recall the terms used in their classification.

Animals are primarily classed in great divisions, based on the general characters which differentiate them most. These divisions, or *branches*, allow of their being so grouped that in each of them we find united the individuals whose general structure is uniform; and under the name of vertebrates are included man and the animals with which our studies will be occupied. The vertebrates, as the name indicates, are recognised by the presence of an interior skeleton formed by a central axis, the vertebral column, round which the other parts of the skeleton are arranged.

The vertebrate branch is divided into classes: fishes, amphibians or batrachians, reptiles, birds, and mammals.

The mammals—from the Latin *mamma*, a breast—are characterized by the presence of breasts designed for the alimentation of their young. Their bodies are covered with hair, hence the name *pilifères* proposed by Blainville; and, notwithstanding that in some individuals the hairs are few, the character is sufficient to distinguish them from all other vertebrates.

We find united in this class animals which, at first, seem out of place, such as the whale and the bat; and, from their external appearance alone, the former would appear to belong to the fishes, and the latter to birds. Yet, on studying their structure, we find that, not only do these animals merit a place in the class which they occupy, because they possess the distinctive characters of mammals; but, still further, their internal structure is analogous to that of man and of the other individuals of this class.

Notwithstanding this similarity of structure, the whale is not without some points of difference from its neighbours the horse and the dog; therefore, in order to place each of these animals in a position suitable to it, mammals are divided into secondary groups called *orders*. The first of these orders includes, under the name *primates*, man and apes. The latter contain animals which approach birds in certain characters of their organism, forming a link between the latter and mammals.

We find, in studying the regions of the body in some of the vertebrates, that, while they present differences from the corresponding regions of the human body, they also offer most striking analogies. We can, for example, recognise the upper limb of man in the anterior one of quadrupeds, in the wing of the bat, in the paddle of the seal, etc. It is, so to speak, those variations of a great plan which give such a charm to the study of comparative anatomy.

The division of classes into orders, which we have just mentioned, being still too general, it was found necessary to establish subdivisions—more and more specialized—to which the names *families*, *genera*, *species*, and *varieties* were given.

[2] Mathias Duval, 'Précis of Anatomy for the Use of Artists': Paris, 1881. 'Artistic Anatomy of the Human Body,' third edition, plates by Dr. Fau, text with figures by Édouard Cuyer: Paris, 1896. 'Artistic Anatomy of Man,' by J. C. L. Sparkes, second edition, text with 50 plates: Baillière, Tindall and Cox, London, 1900.

CHAPTER I

OSTEOLOGY AND ARTHROLOGY

THE TRUNK

The Vertebral Column

We commence the study of the skeleton with a description of the trunk.

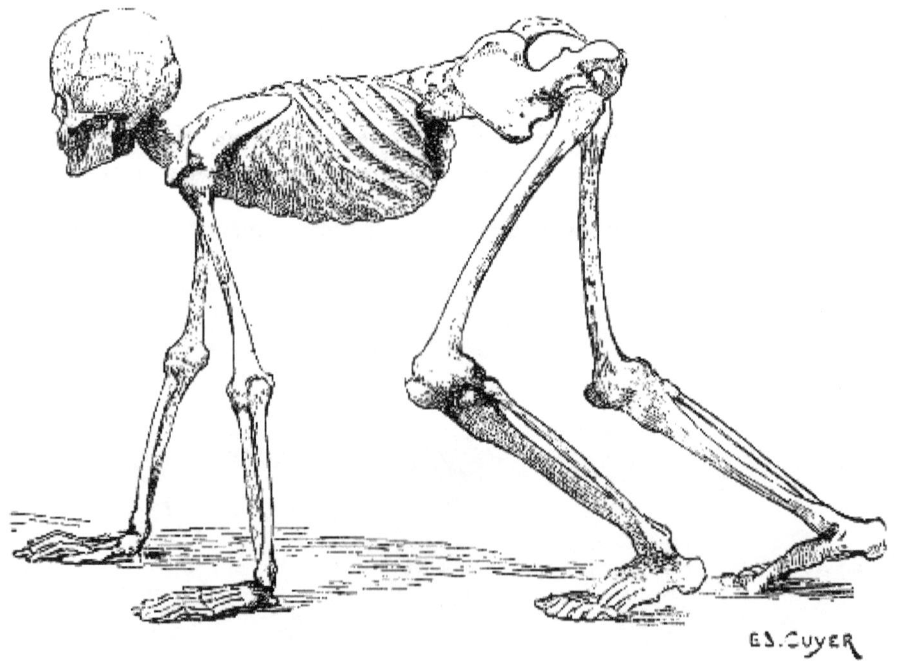

Fig. 1.—A Human Skeleton in the Attitude of a Quadruped. To give a general Idea of the position of the Bones in other Vertebrates.

The trunk being, in quadrupeds, horizontal in direction (Fig. 1), the two regions of which it consists occupy, for this reason, the following positions: the thorax occupies the anterior part, the abdomen is placed behind it; the vertebral column is horizontal, and is situated at the superior aspect of the trunk; it projects beyond the latter: anteriorly, to articulate with the skull; and, posteriorly, to form the skeleton of the tail, or caudal appendix.

The number of the vertebræ is not the same in all mammalia. Of the several regions of the vertebral column, the cervical shows the greatest uniformity in the number of the vertebræ of which it consists, with but two exceptions (eight or nine in the three-toed sloth, and six in the manatee); we always find seven cervical vertebræ, whatever the length of the neck of the animal. There are no more than seven vertebræ in the long neck of the giraffe, but they are very long ones; and not less than seven in the very short neck of the dolphin, in which they are reduced to mere plates of bone not thicker than sheets of cardboard. If the cervical region presents uniformity in the number of its bones, it is not so with the other regions of the column.

The following table shows their classification in some animals:

Vertebræ.

	Cervical.	Dorsal.	Lumbar.
Bear	7	14	6
Dog	7	13	7
Cat	7	13	7
Rabbit	7	12	7
Pig	7	14	6 or 7
Horse	7	18	6 or 5
Ass	7	18	5
Camel	7	12	7
Giraffe	7	14	5
Ox	7	13	6
Sheep	7	13	6

It is worthy of notice that in birds the number of the cervical vertebræ is not constant, as in mammals; they are more numerous than the dorsal. These latter are almost always joined to one another by a fusion of their spinous processes; the two or three last vertebræ are similarly united to the iliac bones, between which they are fixed. The dorsal vertebræ thus form one piece, which gives solidity to the trunk, and provides a base of support to the wings, for the movements of flying. There are, so to speak, no lumbar vertebræ, the bones of that region, which cannot be differentiated from the sacrum, having coalesced with the bones of the pelvis.

Vertebræ.

	Cervical.	Dorsal.
Vulture	15	7
Eagle	13	9
Cock	14	7
Ostrich	18	9
Swan	23	10
Goose	18	9
Duck	15	9

In reptiles, the relation between the number of the cervical vertebræ and that of the dorsal is very variable; some serpents are devoid of cervical vertebræ, having only dorsal ones—that is, vertebræ carrying well-developed ribs.

Vertebræ.

	Cervical.	Dorsal.	Lumbar.
Crocodile	7	14	3
Caiman	7	12	5
Boa	3	248	0
Python	0	320	0
Viper	2	145	0

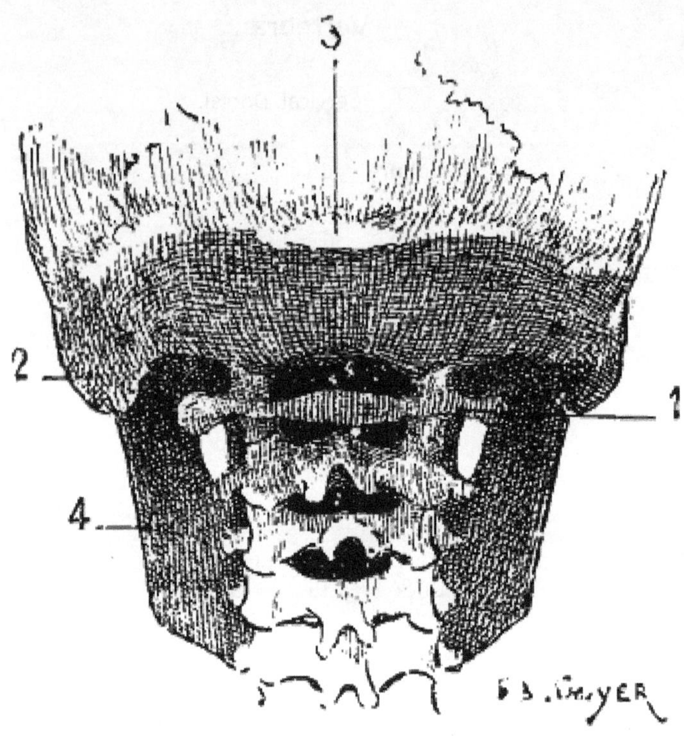

Fig. 2.—Size of the Atlas compared with the Transverse Dimensions of the Corresponding Parts of the Skull in Man.

1, Atlas; 2, mastoid process; 3, external occipital protuberance; 4, inferior maxilla.

Regarding the direction of the vertebral column in animals, in which the trunk is not vertical, it is evident that the spinous processes point upward, and that in comparing them with those of man they must be arranged so that the superior surface of the human vertebra will correspond to the anterior surface of that of the quadruped. Of the cervical vertebræ, the atlas and axis call for special notice. Apropos of the atlas, we find that it, in the human being, is narrower than the corresponding parts of the skull, and is therefore hidden under the base of the cranium (Fig. 2); in quadrupeds its width is equal to that of the skull, and sometimes exceeds, because of the great development of its wing-shaped transverse processes, that of the neighbouring parts of the head (Fig. 3). On this account those transverse processes often project under the skin of the lateral surfaces of the upper part of the neck.

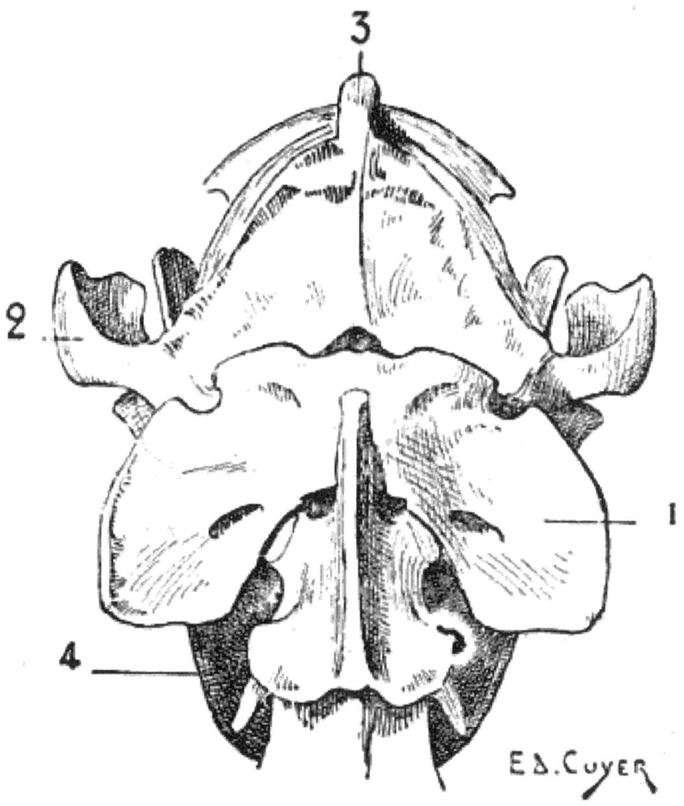

Fig. 3.—Size of the Atlas compared with the Transverse Dimensions of the corresponding Regions of the Skull in a Dog.

1, Atlas; 2, zygomatic arch; 3, external occipital protuberance; 4, inferior maxilla.

The axis is furnished on its anterior surface with the odontoid process, which articulates with the anterior (or inferior) arch of the atlas, according to the direction of the neck. The spinous process, flattened from without inwards, is more or less pointed; it is elongated from before backwards, so as partly to overlap the atlas and the third cervical vertebra.

We find that this process overlaps less and less the neighbouring vertebræ when we examine in succession the bear, the cat, the dog, the ox, and the horse. With regard to the other vertebræ of this region, they diminish in width from the second to the seventh; and, in some animals, the anterior surface of the body presents a tubercle which articulates with a cavity hollowed in the posterior surface of that of the vertebra before it; this feature dwindles away in the dorsal and lumbar regions.

The spinous process, slightly developed in the third cervical vertebra, gradually increases in size to the seventh, the spinous process of which, long and pointed, well deserves the name of *the prominent* which is bestowed on it; but it should not be forgotten that the spinous process of the axis is equally developed.

On the inferior surface of the body of each of the vertebræ is found a prominent crest, especially well marked at the posterior part; this crest is but slightly developed in the bear and in the cat tribe, and is not found in swine.

The transverse processes of the cervical vertebræ, from their relation to the trachea, are known as the *tracheal processes*.

The most marked characteristic of the dorsal vertebræ is furnished by the spinous processes. They are long and narrow. As a rule, the spinous processes of the foremost dorsal vertebræ are the most developed and are directed obliquely upwards and backwards. As we approach the last vertebræ of this region, the processes become shorter and tend to become vertical, and the last ones are even, in some cases, directed upwards and forwards; this disposition is well marked in the dog and the cat. In the cetaceans, on the contrary, the length of the spinous processes increases from the first to the last.

In the horse the spinous processes of the first dorsal vertebræ produce the prominence at the anterior limit of the trunk, where the mane ends, which is known as the *withers*.

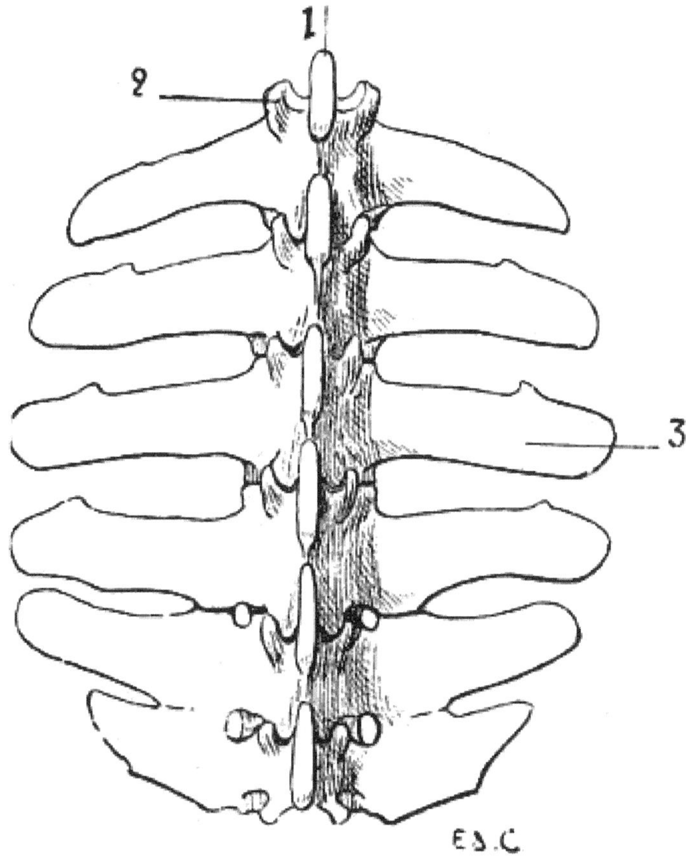

Fig. 4.—Lumbar Vertebræ of a Quadruped (the Horse): Superior Surface.

1, Spinous process; 2, anterior articular process and transverse process of the first lumbar vertebra of the left side; 3, costiform process.

The lumbar vertebræ are thicker than the preceding; they are known by their short and latterly-flattened spinous processes, and still more readily by their transverse processes, which, as they are evidently atrophied ribs, it is more accurate to denominate costiform processes (Fig. 4). These are long, flattened from above downwards, and directed outwards and forwards.

The true transverse processes are represented by tubercles situated on the superior borders of the articular processes of each of the vertebræ of the lumbar region. Apropos of these different osseous processes, we are reminded that they are also present in the human skeleton.

In the horse the costiform processes of the fifth and sixth lumbar vertebræ articulate, and are sometimes ankylosed, one with the other; the terminal ones articulate with the base of the sacrum. Sometimes the processes of the fourth and fifth are thus related; this is the case in the figure (4) given; here the costiform processes of the fourth and fifth vertebræ articulate, and the two terminal ones have coalesced.

In the ox, the same processes are more developed than in the horse; their summits elevating the skin, produce, especially in animals which have not much flesh, prominences which limit the flanks in the superior aspect. The costiform processes of the last lumbar vertebræ are separate from each other; those of the latter are not in contact with the sacrum.

The Sacrum.[3]—This bone, single and median, is formed by the mutual coalescence of several vertebræ, which vary in number according to the species observed.

[3] In human anatomy, the sacrum and the coccyx are studied as part of the pelvis; we, therefore, in the study of the artistic anatomy of man, study these bones with the bones of the lower limbs. Here we do not follow this plan. In animals the sacrum and the coccyx, as a matter of fact, clearly continue the superior border of the skeleton of the trunk; hence we study them with the vertebral column.

Vertebræ Constituting the Sacrum.—Bears, 5; dogs, 3; cats, 3; rabbits, 4; swine, 4; horses, 5; camels, 4; oxen, 5; sheep, 4.

The sacrum is situated between the two iliac bones; with which it articulates, and contributes to the formation of the pelvis. It is obliquely placed, from before backwards, and from below upwards; immediately behind the lumbar section of the vertebral column; and is continued by the coccygeal vertebræ, which form the skeleton of the tail.

It is triangular in outline, and is generally more narrow in proportion than in the human being. All things considered, it is more large and massive, and of greater density, in species which sometimes assume the upright posture, rather than in those which cannot assume that attitude; for example, the sacrum of the ape, of the bear, of the dog, and of the opossum are proportionately larger than those of the horse.[4]

[4] This is particularly striking only in those portions of the sacrum that are not in relation with the other bones of the pelvis. We think that the general form of this bone depends on the mode of its connexions with the iliac bones and the extent of the articular surfaces by which it is in contact with the latter.

Its superior surface presents a crest, formed by the fusion of the spinous processes of the vertebræ which form it. In certain species these processes are attached only by their bases, and are separated from each other superiorly. In the pig they are wholly wanting.

The Coccygeal Vertebræ.—These vertebræ, few in number (and sometimes ankylosed) in the human being, form in the latter a small series, the coccyx; which is inclined forwards, that is to say, towards the interior of the pelvis. In quadrupeds, on the contrary, their number is large; they are not ankylosed, and they form the skeleton of the caudal appendix.

The first coccygeal vertebræ—that is, those which are next the sacrum—present characters which are common to those of other regions: they have a body, a foramen, and processes. As we trace them backwards, these characters become gradually effaced; and they become little more than small osseous cylinders simply expanded at their extremities.

Direction and Form of the Spinal Column

The curves of the vertebral column are, in quadrupeds, slightly different from those which characterize the human spine. First, instead of their being, as in the latter, curves in the antero-posterior aspect, because of the general attitude of the body, they are turned in the supero-inferior direction.

The cervical region is not a single curve, as in the human being. It presents two: one superior, with its convexity looking upwards; the other inferior, the convexity of which is turned downwards. This arrangement reminds one of that of a console.

The dorsal and lumbar regions are placed in a single curved line, more or less concave downwards; so that in the lumbar region there is no curve analogous to that which exists in man; a form which, in the latter, is due to the biped attitude—that is to say, the vertical position of the trunk. Briefly, there is in quadrupeds one dorso-lumbar curve; and not both a dorsal and a lumbar, with convexities in opposite directions.

At the extremity of the dorso-lumbar region is the sacrum and the caudal appendix, which describe a curve of which the concavity is directed downwards and forwards.

It is necessary to point out that it is not the curves of the three anterior portions of the spinal column which determine the form of the superior border of the neck and shoulders, and of the same part of the trunk. For the first portion, there is a ligament which surmounts the cervical region, and substitutes its modelling influence for that of the vertebræ. It is the *superior cervical ligament*, which arises from the spinous process of the first cervical vertebræ, and is inserted into the external occipital protuberance on the upper part of the posterior surface of the skull. The summits of the spinous processes of the vertebræ alone give form to the superior median border of the trunk. In this connection we here repeat that it is not the general curvature of the vertebral column which produces the withers, but the great length of the spinous process of the first vertebræ of the dorsal region.

The Thorax

The dorsal vertebræ form the posterior limit in man, and superior in quadrupeds, of the region of the trunk known as the *thorax*. A single bone, the sternum, is situated at the aspect opposite; the ribs bound the thorax on its sides.

In its general outlines the thorax in quadrupeds resembles that of man—that is to say, that, as in the latter, the anterior portion—superior in the human being—is narrower than the part opposite. But the progressive widening takes place in a more regular and continuous fashion, so that it presents a more definitely conical outline. This purely conical form is nevertheless found in the human species, but only during infancy; the inferior portion of the thoracic cage being then widely expanded, because of the development of the abdominal viscera, which at that period are relatively large.

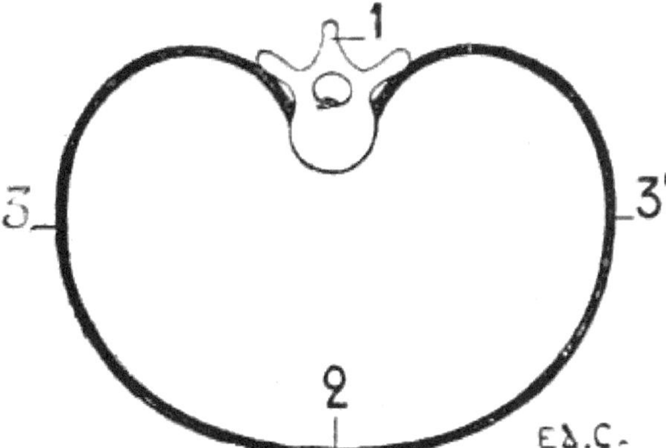

Fig. 5.—A Transverse Section of the Thorax of a Man placed Vertically—that is to say, in the Direction which it would assume in a Man placed in the Attitude of a Quadruped (a Diagrammatic Figure).

1, Dorsal vertebra; 2, sternal region; 3, costal region of one side; 3′, costal region of the other side.

But the proportionate measurements of the thorax are different. Indeed, we may recall that in man the thorax is flattened from before backwards, so that the distance between the sternum and the vertebral column is shorter than the distance from the rib of one side to the corresponding one of the opposite side (Fig. 5). In animals, on the contrary, it is flattened laterally. Its vertical diameter—measured from the sternum to the vertebral column—is greater than the transverse measurement (Fig. 6).

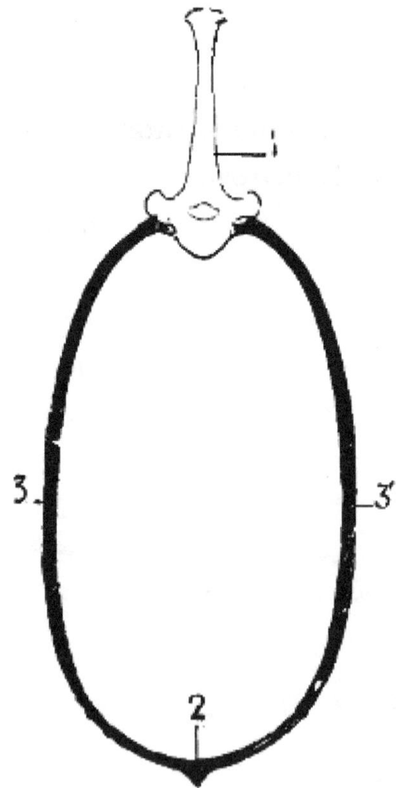

Fig. 6.—A Vertical Section of the Thorax of a Quadruped (Diagrammatic).

1, Fifth dorsal vertebra; 2, sternal region; 3, costal region of one side; 3′, costal region of the opposite side.

From this results a peculiar arrangement of the muscles that we are able to bring directly into prominence, which presents points of interest from the point of view of the contraction of the subcutaneous layer.

Indeed, in man the region occupied by the pectorals is very broad; it is a wide surface turned directly forward. In quadrupeds, this region of the pectorals is narrowed. It is not spread out, as in the preceding instances; and the appearance it presents is explained by the fact that the thorax is compressed laterally. If we examine the thorax on one of its lateral surfaces, the muscles, on the contrary, are more extended. We see the contour of the vertebral column, and the median part of the abdomen; and, especially in the horse, between the great dorsal and the great oblique of the abdomen, we find a large space, in which the ribs, with the intercostals which join them, are uncovered; the muscles in question separate the one from the other, under the influence, it would seem, of the great dimensions of the lateral wall of the thorax.

The Sternum.—The sternum is, in quadrupeds, directed obliquely downwards and backwards; its form varies in different species. In the carnivora, it consists of eight bones, irregularly cylindrical in form, being slightly flattened from within outwards, and thickened at their extremities. They remain separate, and this contributes elasticity and flexibility to the thorax. The first nine costal cartilages articulate directly with the sternum. The first of these cartilages articulates with a nodule situated a little above the middle of the first bone of the sternum.

In the horse the sternum is flattened laterally in its anterior portion, and from above downwards in its posterior half. The six bones which form the sternum are connected by cartilage. The keel-shaped piece, situated in front of the sternum, is also cartilaginous. This process, but slightly marked posteriorly, becomes more and more prominent in front, and terminates at its anterior extremity by a prolongation, slightly curved backwards, which projects for some centimetres beyond the cavity in which the first costal cartilage is received. This process is known as the *tracheal process*, or *rostral cartilage*. The posterior extremity of the sternum, flattened from above downwards, ends in a cartilaginous plate; concave superiorly, and convex inferiorly: this is the abdominal prolongation, or *xiphoid appendix*.

In the ox, the sternum is formed of two distinct bones, which are united by an articulation. One, the anterior, is short, and forms the first portion of the sternum; it is slightly flattened from side to side, and vertical in direction. The other, the posterior, is longer, and is formed by the fusion of several small bones; it is placed horizontally, and is flattened from above downwards. At the level of articulation of these two portions, and because of their different directions, the bone is bent. This bend occurs at the point of articulation of the second costal cartilage. On the superior border of the anterior segment the cartilage of the first rib is articulated. The xiphoid appendix, which is cartilaginous, is attached to the extremity of a long process of the last bone of the sternum.

The shape of the anterior extremity of the sternum is influenced by the presence or absence of clavicles. We have seen that in some quadrupeds the clavicles are wanting. In the first case, this extremity is large, and approaches in shape to the corresponding part of the human sternum, which is so clearly designed to give a point of support to the anterior bone of the shoulder. In the second, on the contrary, this extremity is narrow.

The sternum in birds is very different from that in mammalia, which we have been studying. It varies greatly in extent and shape, under the influence of certain conditions. To understand the cause of these variations it is necessary to remember that in man (as, indeed, in other animals; but the example of man, for that which follows, will be more striking, on account of the mobility of his upper limbs) the sternum gives origin to the pectoral muscles, and that these muscles are inserted into other parts of the thoracic limbs, designed by their contraction to draw the arms downwards, forwards, and inwards—that is, when these are in a state of abduction and in a horizontal direction, they draw them towards the anterior surface of the thorax and downwards. Now, this movement is similar to that made by birds during flight. It is necessary to add that, in the latter case, the more the displacement of the upper limbs has of force and extent, the more the pectoral muscles are developed.

For these reasons, birds, in which, during flight, the movements of the thoracic limbs—the wings—are necessarily energetic, present a great development of the pectoral muscles; having consequently, because an extent of surface for the origin of the muscles commensurate with their development is necessary, a very large and peculiarly shaped sternum (Figs. 18, 6; and 21, 6). Indeed, not only is the sternum large, but, further, in order to form a deeper surface, proportionately adapted to the muscles which arise from and cover it, its anterior surface presents, in the median line, a prominent crest known as the *keel*. This prominence forms two lateral fossæ. We cite as examples, the sternum of the eagle, the vulture, the falcon, and the hawk.

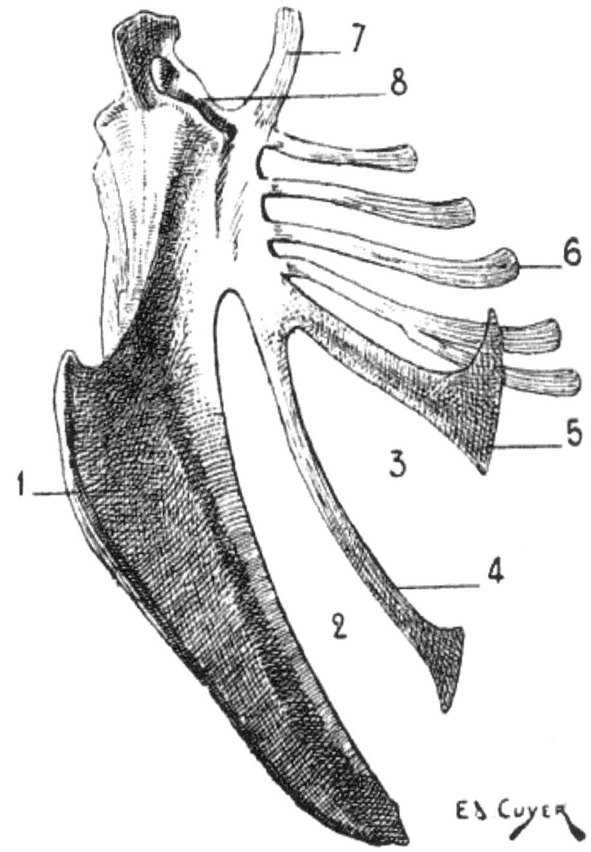

Fig. 7.—Sternum of a Bird (the Cock): Left Side, External Surface.

1, Keel; 2, internal slot; 3, external slot; 4, internal process; 5, external process; 6, inferior ribs; 7, costal process; 8, surface for articulation with the coracoid bone.

All birds are not, however, equally adapted to flight, for in the domestic cock, which flies but a short distance, and badly, the sternum is less developed (Fig. 7); it is also diminished by slots, which diminish its surface. These slots, two on each side, are called from their position the internal and external slots. They are bounded by narrow, elongated, bony processes, an internal and an external; the expanded lower extremity of the latter overlaps the last inferior ribs. The part of the external border which surmounts this external process is hollowed out into grooves, which receive the inferior ribs, and terminates superiorly in an osseous projection known as *the costal prominence*.

In the ostrich, the cassowary, and the apteryx, which run, but do not fly, the sternum has the form of a plate of bone slightly convex, but without a keel.

The shape of the sternum, correlated to the faculty of flight (or of swimming; apropos of which we may cite the penguin, of which the rudimentary wings resemble fins, and perform their functions only), or the absence of this faculty, has furnished the division of birds into two groups. In one are included, under the name *Carinates* (*carina*, keel), those in which the sternum is provided with a keel; in the other division are those in which the sternum is not furnished with one. These latter, on account of their unique mode of progression, are more nearly allied to the mammals.

The keel is developed in flying mammals (bats).

Ribs and Costal Cartilages.—There are on each side of the thorax as many ribs as there are dorsal vertebræ. In animals, as in man, the ribs which articulate with the sternum by their cartilages are called *true*, or *sternal* ribs; those whose cartilages do not articulate with the sternum are called *false*, or *asternal*. The longer ribs are those situated in the middle region of the thorax.

The ribs are directed obliquely downwards and backwards, and this obliquity is more marked in the posterior ones than in the anterior. They are, however, less oblique than in the human being; what proves this is that the first rib in man is oblique, while in quadrupeds it is vertical.

The curvature of the ribs is less pronounced in quadrupeds than in the human being, but this is not equal in all animals. The ribs of the bear are more curved than those of the dog; the latter has ribs more curved than those of the horse.

Each rib, at its vertebral extremity, presents, from within outwards, a wedge-shaped head for articulation with two dorsal vertebræ, a neck, and a tuberosity. External to the tubercle are found some rough impressions, for muscular attachments, which correspond to the angle of the human rib.

In the following table, we give the number and classification of the ribs of some animals:

Number of the Ribs on Each Side of the Thorax.

				Sternal.		Asternal.
Bear	14	divided	into	9	and	5
Dog	13	"	"	9	"	4
Cat	13	"	"	9	"	4
Rabbit	12	"	"	7	"	5
Pig	14	"	"	7	"	7
Horse	18	"	"	8	"	10
Camel	12	"	"	8	"	4
Ox	13	"	"	8	"	5
Sheep	13	"	"	8	"	5

The costal cartilages, by which the first ribs are united to the sternum (sternal ribs), whilst the latter are united one to the other without being directly connected with the sternum (asternal ribs), are, as a rule, in quadrupeds, directed obliquely downwards, forwards, and inwards; each forms, with the rib to which it belongs, an obtuse angle more or less open anteriorly. Their length is proportionate to that of the ribs. The cartilages, which are continued from the asternal ribs, unite and form the borders, directed obliquely downwards and forwards, of the fossa which is found at the inferior and posterior part of the thorax, and which forms the lateral limits of the epigastric region. In the dog and cat the ribs are thick and almost cylindrical; the costal cartilages are thicker at the margin of the sternum than at their costal extremity. In the ox, the ribs are flattened laterally and are very broad, the more so as we examine a portion further from the vertebral column. From the second to the twelfth they are quadrangular in the superior fourth, and thicker than in the rest of their extent. The first costal cartilage is vertical; the following ones are progressively more oblique in a direction downwards and forwards. The four or five cartilages which succeed the first unite with slight obliquity to the sternum; their union with that bone gives the impression of a very strong, well-knit apparatus. The costal cartilages which unite with the sternum are flattened laterally in the portions next the ribs, and flattened from front to back in the rest of their extent.

In the horse the ribs increase in length from the first to the ninth; they are flattened from without inwards, and increase in width from the first to the sixth or seventh, and the following ones become narrower. The costal cartilages, from the second to the eighth, are, as in the ox, at first flattened laterally, near the ribs; while near the sternum they are flattened from front to back.

In birds, the ribs are each furnished with a flat process (Fig. 18, 10), which springs from the posterior border, is directed backwards, and overlaps the external surface of the succeeding rib. These processes are not found, as a rule, on the first or last ribs.

As for the costal cartilages, they are, as a rule, ossified, and receive the name of inferior ribs (Fig. 18, 11), united to the preceding (superior ribs; Fig. 18, 9) by articulation; by the other extremity they are joined to the sternum; the first superior ribs generally want them. Sometimes the last inferior rib becomes connected with the one that precedes it, not articulating with the sternum; and thus recalls the relations of the asternal ribs which we have noticed in our study of the mammals.

In the bat, as in birds, the costal cartilages are ossified.

THE ANTERIOR LIMBS[5]

[5] Consult Figs. 21, 33, 34, 38, 39, 46.

The anterior limbs, homologous to the upper limbs in man, are formed, as in the latter, of four segments: the shoulder, the arm, the forearm, and the hand. These limbs, considered in the vertebral series, present themselves under very different aspects, which are determined by the functions they are called upon to perform.

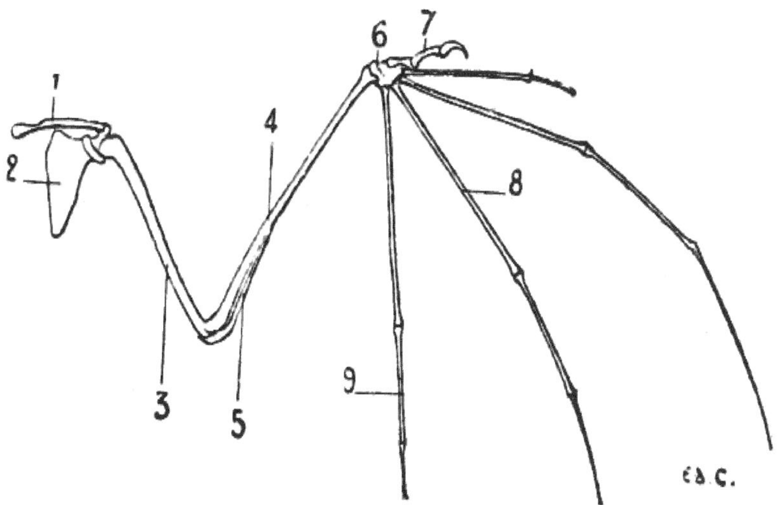

Fig. 8.—Anterior Limb of the Bat: Left Side, Anterior Surface.

1, Clavicle; 2, scapula; 3, humerus; 4, radius; 5, cubitus; 6, carpus; 7, thumb; 8, metacarpus; 9, phalanges.

They constitute the forepaw in terrestrial mammals; in aerial vertebrates they form wings; in aqueous mammals they act as paddles. In whatever series we study them, we can readily find the relationship of the different parts; it is very easy to recognise the same bones in the upper limbs of the human being, the wings of the bat (Fig. 8) and of birds (Fig. 21), and in the anterior paddles of the seal (Fig. 9) and of the dolphin.

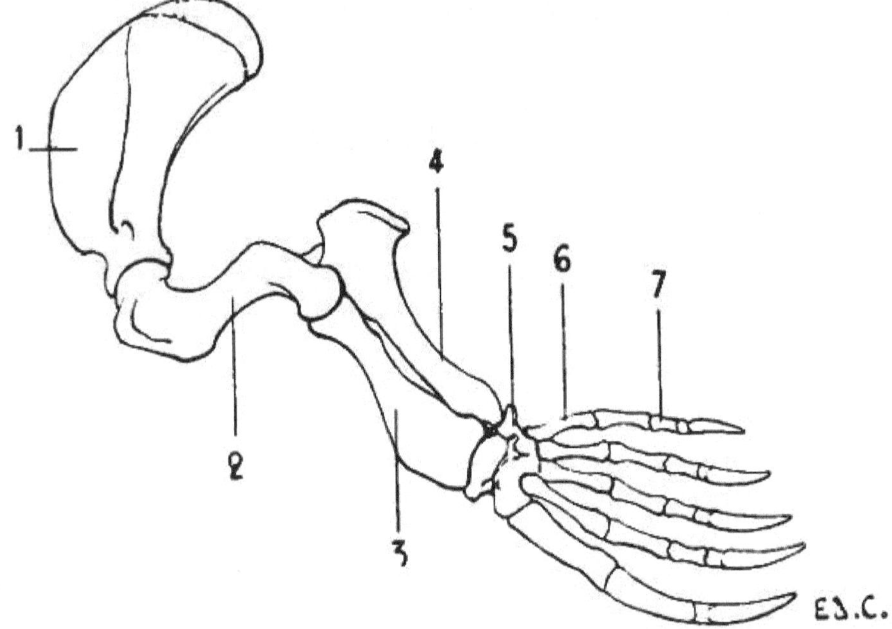

Fig. 9.—Anterior Limb of the Seal: Left Side, External Surface.

1, Scapula; 2, humerus; 3, radius; 4, ulna; 5, carpus; 6, metacarpus; 7, phalanges of the fingers.

In quadrupeds, the shoulder and arm are hidden, the latter more or less completely, in the muscular mass which binds it to the lateral wall of the trunk; so that the anterior limbs only present; free from the trunk: the elbow, forearm, and hand.

The Shoulder

In some vertebrates, the shoulder is formed of two bones—the scapula and clavicle; in others of only one bone—the scapula; the clavicle in this case does not exist.

The Scapula or Omoplate.—The scapula is situated on the lateral surface of the thorax, and is directed obliquely, from above downwards and from behind forwards.

We must first recall, so as to be able to make a comparison, that in man this bone is placed at the posterior surface of the thoracic cage; so that if we look at the human thorax on one of its lateral aspects we see chiefly the external border of the scapula; it is the external surface (homologous to the posterior surface of the human scapula) which we see in its full extent when we look on the same surface of the thorax in quadrupeds.

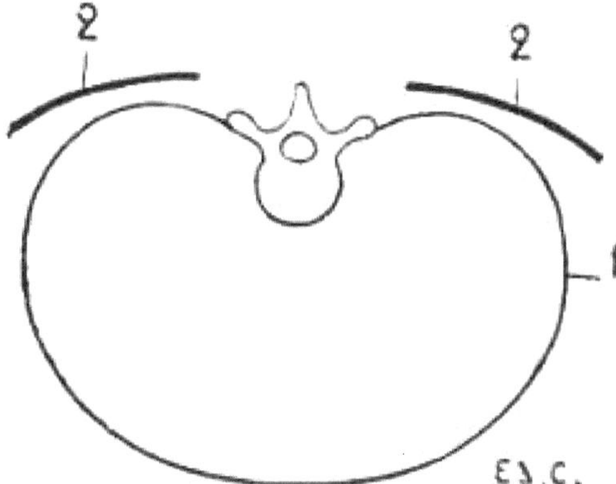

Fig. 10.—Situation and Direction of the Scapula in the Human Being, the Trunk being Horizontal, as in Quadrupeds. Vertical and Transverse Section of the Thorax (Diagrammatic Figure).

1, Contour of the thorax; 2, 2, the scapula.

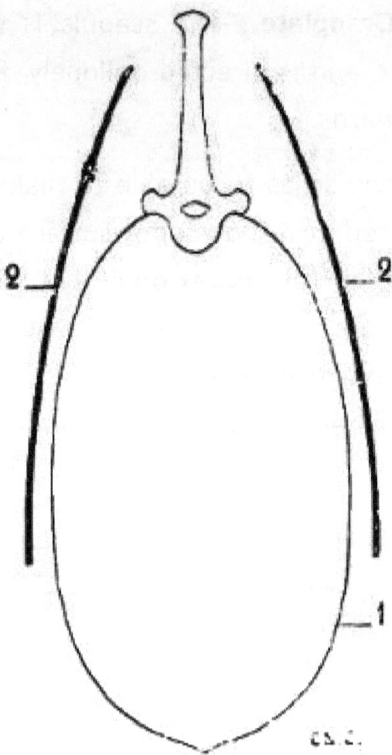

Fig. 11.—Position and Direction of the Scapula in Quadrupeds. Vertical and Transverse Section of the Thorax (Diagrammatic Figure).

1, Contour of the thorax; 2, 2, the scapula.

To sum up, if we fancy the human being in the position of the quadruped, the scapula will have its surfaces almost parallel to the ground (Fig. 10); while in quadrupeds, the surfaces are situated in a plane which is almost perpendicular to the ground (Fig. 11). This position of the scapula in an almost vertical plane is designed to give the necessary point of support to the osseous columns that form the skeleton of the other portions of the anterior limbs.

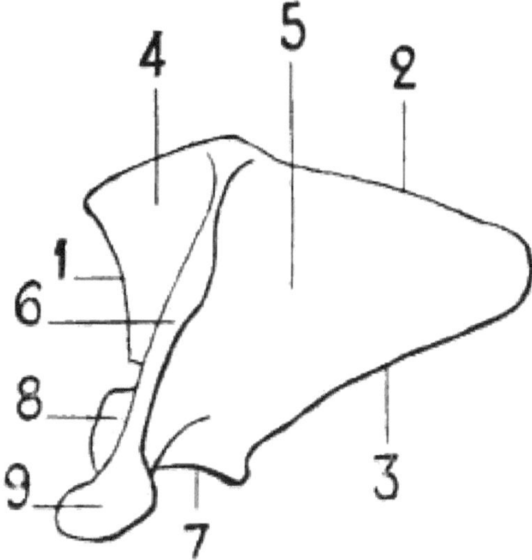

Fig. 12.—Left Scapula of the Human Being, Posterior Surface, placed in the Position which it would Occupy in the Skeleton of a Quadruped.

1, Cervical border; 2, spinal border—the scapula here represented, being from a hoofed animal, has a cartilage of extension attached to its spinal border; 3, axillary border; 4, supraspinous fossa; 5, subspinous fossa; 6, spine of the scapula; 7, glenoid cavity; 8, coracoid process. The scapula of the horse has no acromion process, but it is easy, if we compare the human scapula, to judge of the position which this process would occupy if it were present.

Because of this position of the scapula (Figs. 12 and 13), the spinal border is superior, the cervical, anterior, and the axillary, posterior. In direct contrast to what obtains in the human scapula, the spinal border is the shortest of the three; except in the bat, and the majority of the cetaceans.

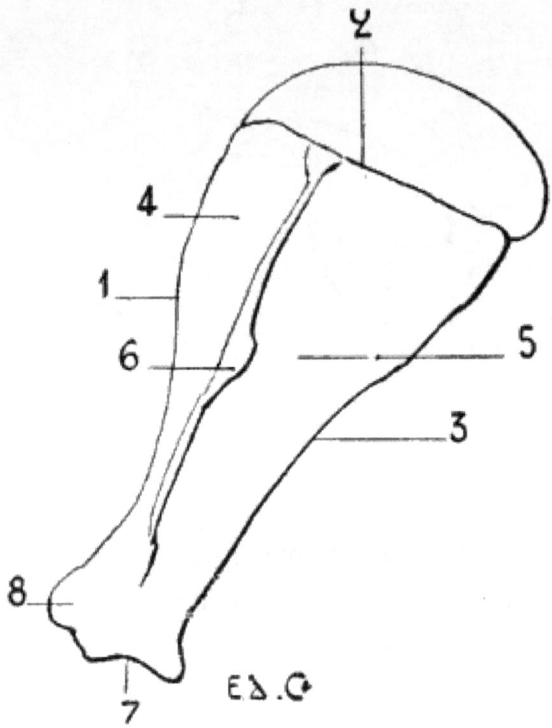

Fig. 13.—Left Scapula of a Horse: External Surface.

1, Cervical border; 2, spinal border; 3, axillary border; 4, supraspinous fossa; 5, subspinous fossa; 6, scapular spine; 7, glenoid cavity; 8, coracoid process; 9, acromion process.

In certain animals (in the ungulates [*hoofed*[6]]—pigs, oxen, sheep, horses) the superior, or spinal, border of the scapula is surmounted by a cartilage called *the cartilage of prolongation*.

[6] For the definition of the word *hoofed*.

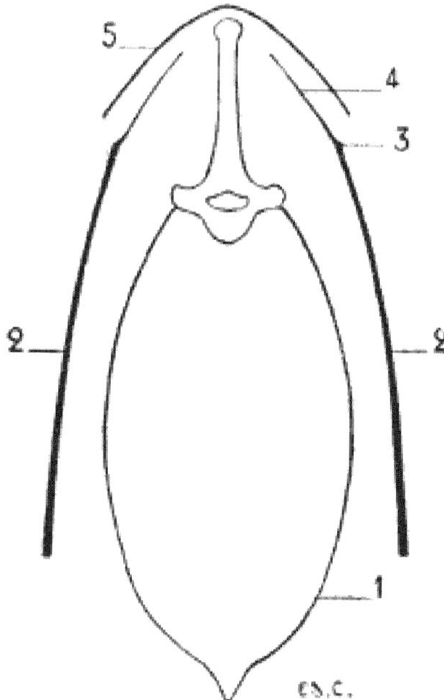

Fig. 14.—Vertical and Transverse Section, at the Site of the Shoulders, of the Thorax of the Horse (Diagrammatic Figure).

1, Outline of the thorax at the level of the third dorsal vertebra; 2, 2, scapula; 3, spinal border of the scapula; 4, cartilage of prolongation; 5, contour of the skin.

This is the cause why the border to which it is fixed is so slightly noticeable under the skin in these animals; indeed, in the upper part, the bone and cartilage are not distinguishable in the contour of the corresponding region of the back; being applied to the lateral surfaces of the spinous processes, the prominence formed by the extremities of which is directly continuous with the plane of the scapula (Fig. 16).

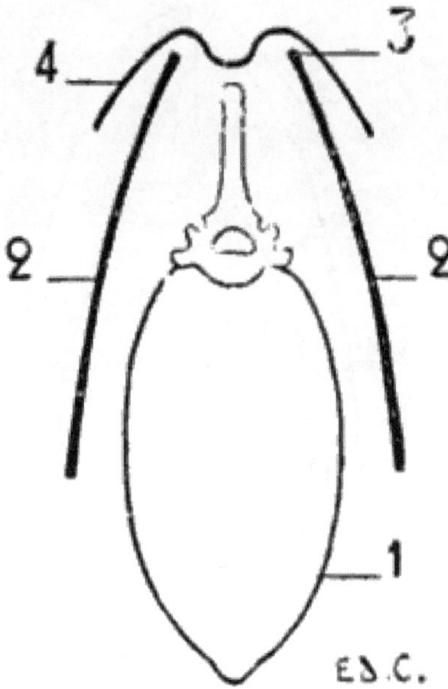

Fig. 15.—Vertical and Transverse Section, at the Plane of the Shoulders, of the Thorax of a Dog (Diagrammatic Figure).

1, Outline of the thorax at the level of the third dorsal vertebra; 2, 2, scapula; 3, spinal border of the scapula; 4, contour of the skin.

In quadrupeds whose scapula, on the contrary, is wanting in the cartilage of prolongation (in the *clawed*,[7] such as the cat and dog), the superior border of the scapula is visible, especially when the animal is resting on its fore-limbs, particularly when it crouches; at such a time the skin is markedly raised by that border; and the spinous processes of the vertebræ, beyond which it projects, occupy the bottom of a fossa (Fig. 15). The internal surface of the scapula is turned towards the ribs; it is known, as in man (in whom this surface is anterior), as the subscapular fossa.

[7] For the definition of this word.

Fig. 16.—Left Clavicle of the Cat: Superior Surface (Natural Size).

1, Internal extremity; 2, external extremity.

Fig. 17.—Clavicle of the Dog (Natural Size).

Its external surface is divided into two parts by the spine of the scapula; which, in some animals, terminates inferiorly in a flat and clearly distinct process, the homologue of the acromion process of the human scapula. The two regions separated by the spine are known as the supraspinous fossa and the infraspinous fossa. The supraspinous fossa is anterior to the spine, and the infraspinous is posterior to it. The surfaces of the scapula are, in quadrupeds, flatter than in the human being, and in particular the subscapular fossa, which is also less concave. Some authors attribute this to the lesser curvature of the ribs in quadrupeds. A few words will suffice to prove that there must be another reason. The scapula is not in immediate contact with the ribs; the subscapular fossa is not moulded on them. Besides, the form of the scapula is, as in other parts of the skeleton, dependent on the disposition of muscles, and the development of these latter is correlated to the extent and energy of the movements which the individual is able or required to execute. But the movements which those muscles produce (more especially the rotation of the humerus) are, in quadrupeds, less extensive than in the human being; and, consequently, the muscles which produce them are, proportionally, less strongly developed. The inferior angle (superior and external in man), situated at the junction of the cervical and axillary borders, presents the glenoid cavity, which, looking downwards, receives the articular surface of the superior extremity of the bone of the arm—that is to say, the head of the humerus. Above this cavity, on the lower part of the cervical border, is situated a tubercle which reminds us of the coracoid process of the human scapula. The region occupied by the glenoid cavity is separated from the body of the bone by a constriction—the neck of the scapula.

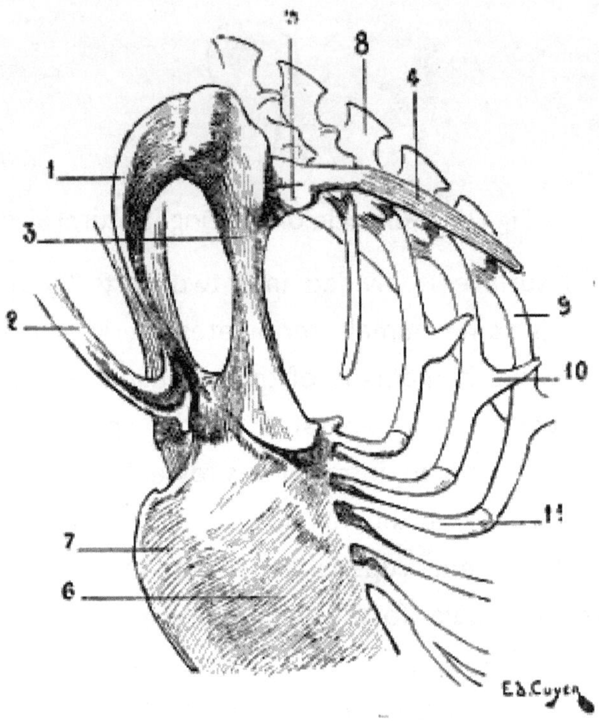

Fig. 18.—Skeleton of the Shoulder of a Bird (Vulture): Antero-External View of the Left Side.

1, Left clavicle; 2, inferior portion of the right clavicle, forming by its ankylosis with that of the other side the fourchette; 3, coracoid bone; 4, scapula; 5, articular surface for humerus; 6, superior half of the sternum; 7, keel of sternum; 8, spinous process of the dorsal vertebræ; 9, superior ribs; 10, process of one of these ribs; 11, inferior ribs.

In birds the scapula is elongated in a direction parallel to the vertebral column, and very narrow in the opposite (Fig. 18): it is also flat, and has no spine. Its coracoid process is represented by a peculiar bone—the coracoidean or coracoid bone—which we shall describe later on when we come to the study of the clavicle and of the anterior region of the shoulder.

The Clavicle.—The clavicle is found only in the human being, and in animals whose anterior limbs, possessing great freedom of movement in all directions, require that the scapula should possess a point of support which, while affording this, can be displaced with it, or draw it in certain directions. Now, this point of support is furnished by the clavicle.

In animals possessed of hoofs (ungulates), such as the sheep, ox, and horse, the clavicle does not exist. Indeed, in them the freedom of movement of the anterior limbs is limited; they move by projection in the forward and backward directions only; they merely fulfil the functions of giving support to and carrying about the body. The clavicle is rudimentary in the cat and the dog; in the cat it is a small, elongated bone (Fig. 16), 2 centimetres in length, thin and curved, connected with the sternum and the scapula by ligamentous bundles. In the dog it is represented by a small osseous plate only (Fig. 17), which is not connected with any of the neighbouring bones.

It is on the deep surface of a muscle which passes from the head and neck to the humerus (mastoido-humeral, a muscle common to the arm, neck, and head) in which this rudimentary bone is found to be developed.

The clavicle exists in perfect state in mammals which use their limbs for digging, grasping, or flying; the insectivora (hedgehog, mole) and some rodents (squirrel, woodchuck) are provided with it.

The cheiroptera (bats) possess an extremely well-developed clavicle, on account of the varied movements which their thoracic limbs execute.

This formation of the shoulder which favours flight in the bat is even more remarkable in birds. In these latter (Fig. 18) the clavicles, fused together by their lower extremities, form one bone, having the shape of the letter V or U, which is known as the *fourchette*; this bone, acting as a true spring, keeps the shoulders apart, and prevents their approximation during the energetic movements which flight necessitates.

In birds whose power of flight is strong, the two limbs of this bone are widely separated and thick, and the fourchette is U-shaped. Those whose flight is awkward and but slightly energetic have the limbs of the fourchette slender; they unite at a more acute angle, and the bone is V shaped.

Furthermore, a bone named the *coracoid* joins the scapula to the sternum; this bone, often fused with the scapula, where it contributes to the formation of the glenoid cavity, represents in birds the coracoid process of the human scapula. If we fancy this process directed inwards, and sufficiently lengthened to join the sternum, we shall have an idea of the disposition of the bone we are now discussing, and the reasons for which the name has been chosen by which it is designated. The coracoid bone, like the fourchette which it reinforces, offers to the wings a degree of support proportionate to the efforts developed by those limbs; for this reason it is thick and solid in birds of powerful flight.

The superior extremity of each branch of the fourchette, at the level of its junction with the coracoid and the scapula, bounds, with these latter, a foramen which gives passage to the tendon of the elevator muscle of the wing, or small pectoral. The importance of the fourchette being, as we have seen, in proportion to the movements of flying, it is easy to understand that the bone is not found in the ostrich.

The Arm

A single bone, the humerus, forms the skeleton of this portion of the thoracic limb.

The Humerus.—The bone of the arm is, in quadrupeds, inclined from above downwards and from before backwards.

It is, with relation to other regions, short in proportion as the metacarpus is elongated, and as the number of digits is lessened. In the horse, for example, whose metacarpus is long, and in which but one digit is apparent, the humerus is very short. The slight development in length of the humerus explains its close application to the side of the animal as far as the elbow.

In animals in which the humerus is longer, the bone is slightly free, as well as the elbow, at its inferior extremity. Later on we will return to the consideration of this peculiarity and of the proportions of the humerus, after we have studied the other parts of the fore-limbs.

The humerus in quadrupeds is inflected like the letter S; in man this general form is less accentuated, the humerus being almost straight. On its body, which appears twisted on its own axis, we find the musculo-spiral groove,[8] which crosses the external surface, and is very deep in some animals. Above this groove, and on the external surface, there exists a rough surface which is the impression of the deltoid. In some species this rugosity is very prominent, and is called *the tuberosity of the deltoid*; it is prolonged downwards by a border which forms the anterior crest of the musculo-spiral groove and limits this latter in front. The external border of the bone, or posterior crest of the groove, limits it behind.

[8] It would be going outside our province to discuss whether the humerus is really twisted on its axis. This question, often discussed, has been solved in some recent works in the following manner: the humerus has undergone torsion at the level of its superior extremity, and not at the level of its body; this does not authorize us further to accord any definite sense to the denomination 'groove of torsion' (musculo-spiral groove). That which we must especially remember in connection with this fact, is, as we shall afterwards see, the difference of direction which the articular head presents according as the torsion has been more or less considerable: because this is established, according to the same order, in man and in quadrupeds.

The superior extremity is enlarged, and remarkable in three portions which it presents; these are: an articular surface and two tuberosities.

The articular surface, or head of the humerus, smooth and round, is in contact with the glenoid cavity of the scapula. This head in the human skeleton is directed upwards and inwards; in quadrupeds its direction is upwards and backwards. The inferior extremity, having in both one and the other its long axis directed transversely, and the point of the elbow looking backwards in all, the result is that the head of the humerus is not situated vertically above the same regions; in the first, it is almost directly above the internal part of this extremity; in the latter, it is situated above its posterior surface, or the point of the elbow in the complete skeleton. This difference of direction is correlated with the position of the scapula, the glenoid cavity of which, as we have already seen, is in man turned outwards, whereas in quadrupeds it looks downwards. In the latter case the scapula consequently rests on the head of the humerus; and this position is most favourable for the performance of the functions which the anterior limbs have to fulfil in these latter.

Of the tuberosities of the head of the humerus, one is situated on the external aspect—it is the great tuberosity, or *trochiter*; the other is placed internally—it is the small tuberosity, or *trochin*. The great tuberosity is divided into three parts—summit, convexity, and crest; these different parts give insertion to the muscles of the shoulder. We recollect that the facets (anterior, middle, and posterior) of the great tuberosity of the humerus in man give attachment to the muscles of the same region. The head of the humerus in the human body projects above the tuberosities. We shall see afterwards, when dealing with some special quadrupeds, that in some of these, on the other hand, the tuberosities are on a higher level than the articular head of the bone. Between the two tuberosities is the bicipital groove.

In man, the superior extremity of the humerus, although covered by the deltoid, reveals its presence by elevating the corresponding portion of the latter. In quadrupeds, the anterior part of this extremity, although similarly covered by muscular bundles, produces a prominence under the skin. This prominence is situated at the summit of the angle formed by the opposing directions of the scapula and the bone of the arm, and constitutes what is known by the name of the *point of the shoulder*, or of the *point of the arm*.

The inferior extremity, transversely enlarged, presents an undulating articular surface, which reminds us of the trochlea and the condyle of the human humerus; on which, however, the condyle is more sharply defined from the trochlea.

In the human skeleton, the internal lip of the trochlea descends lower than the external; and also lower than the condyle. In the bear, the cat, and the dog, it is the same. In the ox and the sheep, the condyle is lower than the trochlea, but only very little lower. In the horse the arrangement is still the same, but a little more accentuated.

On the lateral parts of this extremity we find: internally, a prominence, the epitrochlea; and, externally, another, the epicondyle. It is from this latter that the crest arises, which, passing upwards, forms the posterior limit of the groove of torsion.

The two prominences, which we have just described from a general point of view, present special arrangements which it is necessary to point out. When we examine the form of the outline of the inferior extremity of the humerus in man, the bear, the cat, the dog, the ox, and the horse, we find in following this order that the extremity tends to become narrow transversely, and that the epicondyle and the epitrochlea are less and less prominent on the external and internal aspects respectively. These two processes, indeed, project backwards; the epitrochlea always remaining more developed than the epicondyle. Because of this projection backwards, the cavity situated on the posterior surface of the inferior extremity, the olecranon fossa, is very deep, more so than in the humerus of man. Its borders being thus formed by the two processes, are very prominent. In front we find the coronoid fossa, which is less deep than that of which we have just spoken.

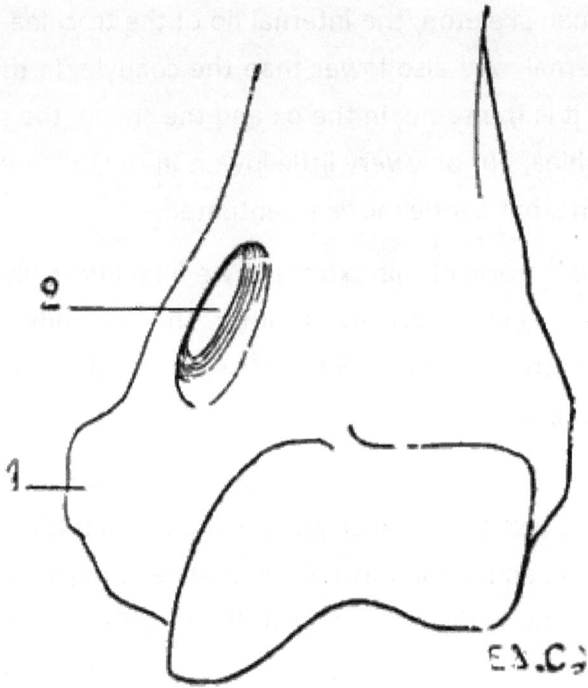

Fig. 19.—Inferior Extremity of the Left Humerus of a Felide (Lion).

1, Epitrochlea; 2, supra-epitrochlear foramen.

There exists in some mammals an osseous canal, situated above the epitrochlea, and known as the *supratrochlear canal* (Fig. 19). It is bounded by a plate of bone which at its middle portion is detached from the shaft of the humerus, and blends with the latter at both its extremities. The brachial artery and median nerve pass through the foramen.

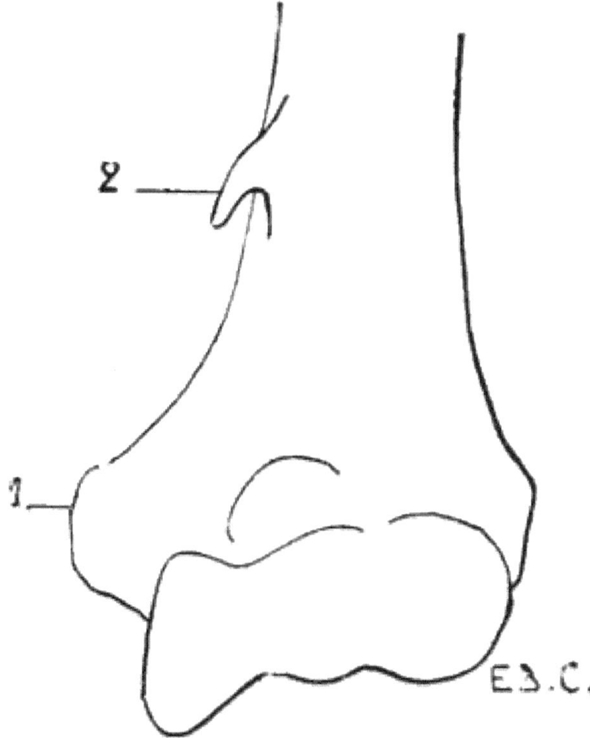

Fig. 20.—Inferior Extremity of the Left Human Humerus, showing the Presence of a Supratrochlear Process.

1, Epitrochlea; 2, supra-epitrochlear process.

A similar condition is sometimes found, as an abnormality, in man, which presents itself under the following aspect (Fig. 20): an osseous prominence more or less long, in the shape of a crochet-needle—supra-epitrochlear process—situated 5 or 6 centimetres above the epitrochlea; the summit of this process gives attachment to a fibrous band, which is inserted by its other end into the epitrochlea and the internal intermuscular aponeurosis. The fibro-osseous ring thus formed gives passage to the brachial artery and the median nerve, or in case of a premature division of this artery to the ulnar branch of the same.[9]

[9] For further details of this anomaly, see Testut, 'The Epitrochlear Process in Man' (*International Journal of Anatomy and Physiology*, 1889); A. Nicolas, 'New Studies on the Supratrochlear Process in Man' (*Review of Biology of the North of France*, t. iii., 1890-1891).

There is also found in some mammals a perforation of the thin plate of bone which, in others, separates the olecranon fossa from the coronoid. This perforation is sometimes found as an abnormality in the human humerus.

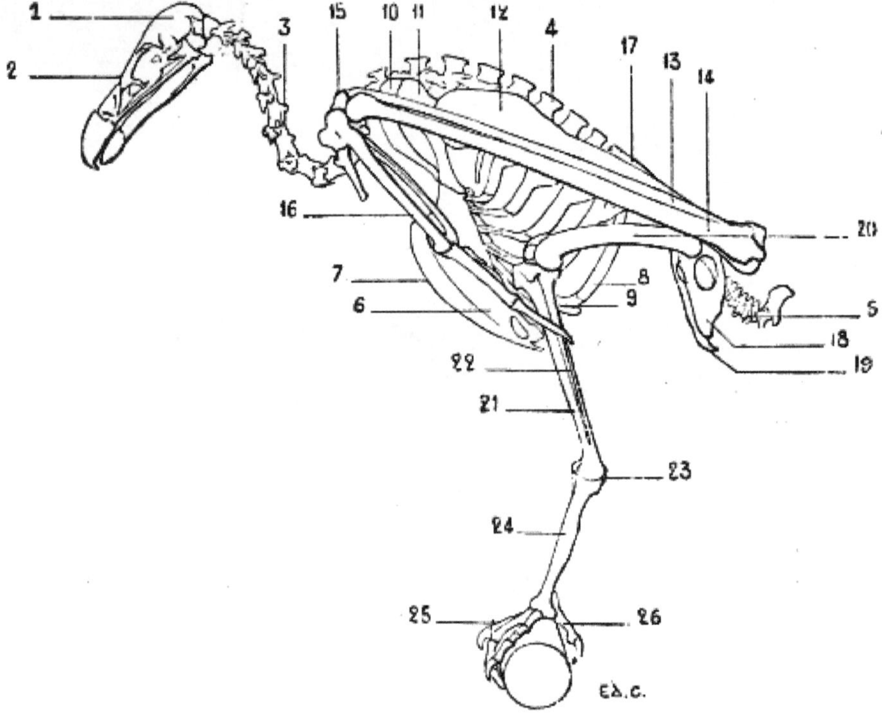

Fig. 21.—Skeleton of a Bird (Vulture): Left Surface.

1, Cranium; 2, face; 3, cervical vertebræ; 4, spinous processes of the dorsal vertebræ; 5, coccygeal vertebræ; 6, sternum; 7, keel; 8, superior ribs; 9, inferior ribs; 10, clavicle; 11, coracoid bone (for the details of the skeleton of the shoulder, see Fig. 18); 12, humerus; 13, radius; 14, ulna; 15, carpus; 16, hand (for details of the skeleton of this region, see Fig. 31); 17, ilium; 18, ischium; 19, pubis (for the details of the pelvis, see Fig. 46); 20, femur; 21, tibia; 22, fibula; 23, osseous nodule, which some anatomists think represents the calcaneum; it is the sole vestige of the tarsus; 24, metatarsus; 25, foot; 26, first toe (for the details of the skeleton of the foot, see Fig. 48).

As does the sternum and the skeleton of the shoulder, the humerus of birds presents differences correlated to the functions which the thoracic limbs are destined to fulfil. Lying on the side of the thorax, directed obliquely downwards and backwards (Fig. 21), it is proportionately longer in individuals of powerful flight than in those which fly less or not at all. In the vulture it projects beyond the posterior part of the pelvis; in the cock it does not even reach the anterior border of the same. To these differences in length are added differences in volume and in the development of the processes which serve for muscular attachment, which are more considerable in birds of powerful flight.

The humerus is so placed that the radial border, external in man and quadrupeds, looks upwards, with the result that the surface of the bone of the arm, which in these latter is anterior, in the former looks outwards. The humeral head, which is turned forwards and a little inwards, is convex and elongated in the vertical direction. Behind and above this head is found a crest for the insertion of muscles. It is the same for the region below, where there is a tuberosity whose inferior surface presents a pretty large opening which looks inwards to a fossa from the floor of which a number of minute openings communicate with the interior of the bone. This is the pneumatic foramen of the humerus.

It is of interest to remember in connection with this subject that in birds, in keeping with the conditions of flight, every system of organs is adapted to diminish the weight of the body. We particularly draw attention to the osseous framework, the structure of which is such that the weight of the animal is greatly lessened. This condition is secured by the pneumaticity. The bone consists of a cover of compact tissue, which, instead of enclosing marrow, is hollowed out by cavities which contain air, and communicate with special pouches, the air-sacs, which are appendages of the lungs.[10]

[10] The presence of air in the bones does not seem to be always associated with the power of flight; as a matter of fact, we find air spaces in the bones of some birds which do not fly (E. J. Marey, 'The Flight of Birds,' Paris, 1890, p. 51).

The antibrachial extremity of the humerus is flattened from without inwards. It terminates in two articular surfaces, which articulate with the radius and ulna.

The olecranon process of the ulna being slightly developed, it follows that the olecranon fossa is not large; neither is the coronoid.

General View of the Form of the Forearm and Hand

We now proceed to the study of the two regions of the fore-limbs which present the greatest variety in regard to the number of bones and also in regard to form and proportions. These two regions are the forearm and the hand.

It is first of all necessary to say that in man, when the fore-limb hangs beside the body, and the dorsum of the hand looks backwards, the two bones of the forearm are parallel, and that this position is known by the name of *supination*. It is also necessary to remember that there is another attitude, in which the radius, crossing the ulna, and carrying the hand with it, displaces the latter in such a way that the palmar surface looks backwards. This second position is known as *pronation*.

Let us now suppose that a man wishes to walk in the attitude of a quadruped. It will be necessary, in order that his upper limbs, being for the moment anterior ones, may act as members of support, to place the forearm in pronation, in order that, as is more normal, the hands may rest on the ground by their palmar surfaces. In this position the radius, being rotated on its own axis at its upper extremity and around the ulna in the rest of its extent, shall have its inferior extremity situated on the inner side of the corresponding extremity of the latter.

Such is the situation of the bones of the forearm and the attitude of the hand in quadrupeds. In short, quadrupeds have their anterior members in the position of pronation.

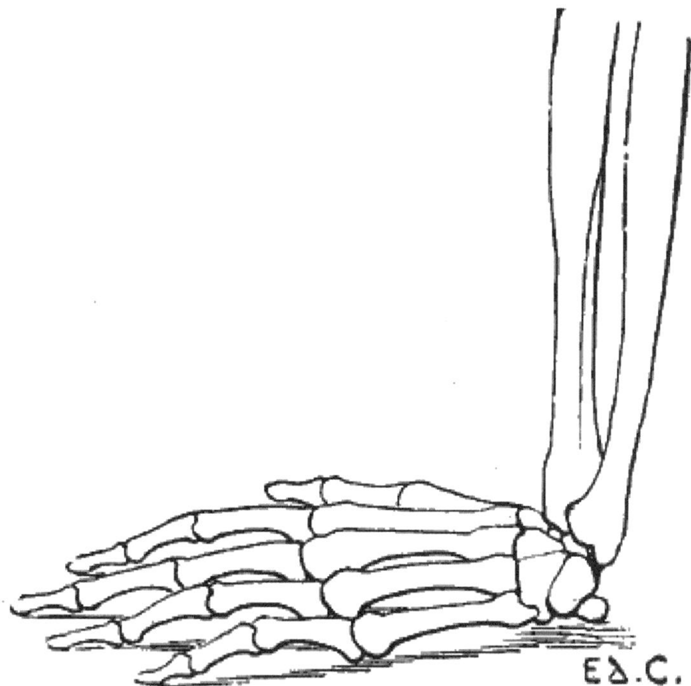

Fig. 22.—The Human Hand resting for its Whole Extent on its Palmar Surface: Left Side, External Surface.

The individual whom we have just supposed placed in the attitude of a quadruped would be able to maintain this position by pressing on the ground more or less extensive portions of his hands; the whole palm of the hand may be applied to the ground (Fig. 22); or the fingers only—that is to say, the phalanges (Fig. 23); or the extremities of the fingers only—that is to say, the third phalanges (Fig. 24). This last position, which is certainly difficult to maintain, should here be regarded rather as theoretical.

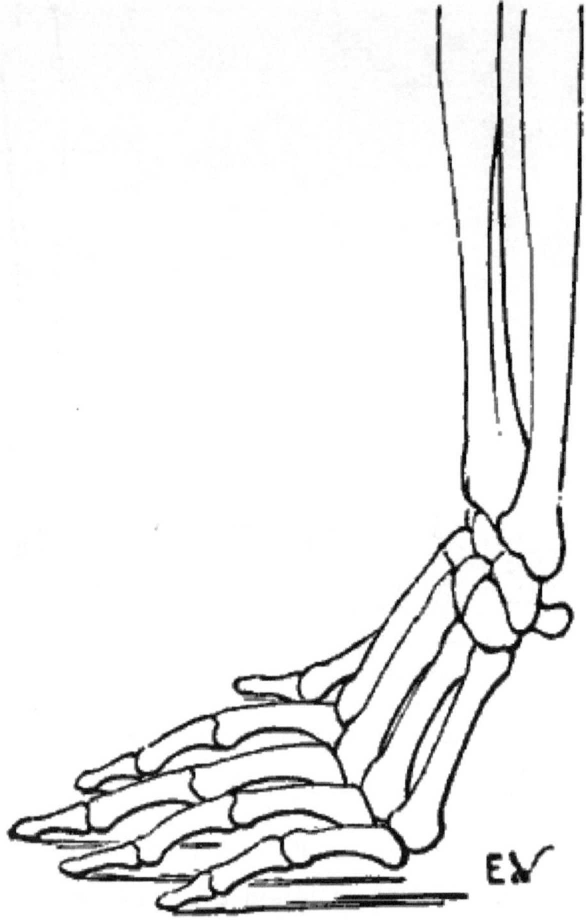

Fig. 23.—The Human Hand resting on its Phalanges: Left Side, External Surface.

We shall meet with each of these modes of support in certain groups of animals. Thus, the bear, badger, and the majority of rodents, have the paws applied to the ground by the whole extent of the palmar surface of the hand, from the wrist to the tips of the fingers. They are therefore called plantigrade, from the analogy, in this case, of the palm of the hand to the plantar surface, or sole of the foot.

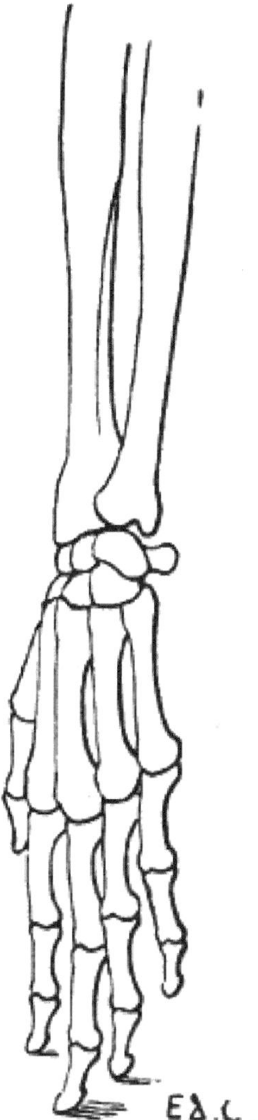

Fig. 24.—The Human Hand resting on the Tips of some of its Third Phalanges: Left Side, External View.

In others, such as the lion, tiger, panther, cat, wolf, and fox, the support is made no longer on the whole extent of the palmar surface, but on the corresponding surface of the fingers only—the metacarpus is turned back, and, consequently, the wrist—that is to say, the carpus—is removed from the ground. These are the digitigrades.

Lastly, the ruminants (sheep, oxen, deer, etc.), and also the pig, ass, and horse, rest on the third phalanx only. In them not only is the metacarpus turned back, but also the two first phalanges. The wrist is very far removed from the ground. In these animals, the third phalanx is enclosed in a case of horn, a nail (the hoof), and because the support of the limb is on that nail, the name of unguligrades has been given them. Nevertheless, as the point of support is on the third phalanx, which is also known by the name of phalangette, we are of opinion that, in order to specify definitely, although they walk on their fingers, as do the digitigrades, the support is provided not by the whole extent of those appendages, they might receive the name of phalangettigrades.

It is necessary among the ruminants to make an exception of the camel and the llama, which are digitigrades.

Just in proportion as the hand is raised from the ground, as we have just seen in passing from the plantigrades to the digitigrades and unguligrades, the number of bones of that region diminishes, the bones of the forearm coalesce, and the ulna tends to disappear; the hand becomes less and less suitable for grasping, climbing, or digging, so as to form an organ exclusively adapted for walking and supporting the body.

Thus, the bear (plantigrade) has five digits, and the power of performing the movements of supination and pronation. Indeed, we know with what facility this animal is able to move his paws in every direction, and climb a tree by grasping it with his fore-limbs. It is well known, however, that no animal except the ape can perform the movements of rotation of the radius around the ulna with the same facility as man; and that none possesses the same degree of suppleness, extent, and variety of movements of the forearm and hand.

In the digitigrades there is one finger which is but slightly developed, and which is always removed from the ground—that is, the thumb: there is also a little less mobility of the radius around the ulna.

In the ungulates the limbs are simply required to perform the movements of walking, and form veritable columns of support, which become the more solid as they are less divided. The bones of the forearm are fused together; there is therefore no possibility of rotation of the radius around the ulna. The metacarpus is reduced to a single piece, which in the horse constitutes what is known as the *canon*. The number of digits becomes diminished, so that in ruminants there are not more than two, and in the horse but one. We should, however, add that, up to the present, we have taken into account only perfect digits, those that rest on the ground. We shall see further on that there exist supplementary digits, but that they are only slightly developed, and are represented in some cases by mere osseous spurs; it is this fact that has permitted us to ignore them in the general study which we have just made.

Because, as we have already said, the unguligrades have the inferior extremity of the digit encased in a horny sheath, which forms the hoof of the horse and the corresponding structures (*onglons*) in the ox, those animals have been placed in a special group, which is based on that peculiarity—that is, the group of ungulate mammals.

The plantigrades and digitigrades, of which the paws have their surfaces of support strengthened by an epidermic sole and fatty pads, have the free extremities of the third phalanges covered on their dorsal surface by nails or claws; hence they are named *unguiculate* mammals.

The bat and birds have the bones of the forearm so arranged that the radius cannot rotate around the ulna. This is necessary in order that during flight, when the wing is being lowered, the radius and hand shall not be able to turn; for, if such rotation took place, each stroke of the wing would place it in a vertical position, which would occasion a loss of resistance incompatible with the effect to be obtained.

The Forearm

The skeleton of the forearm in quadrupeds is vertical in direction; consequently, it forms with the arm an angle open anteriorly; this is well seen on examining the lateral surface. If we examine it on its anterior surface, we find a slight obliquity directed downwards and inwards. In animals in which the bones of the forearm are separate—that is to say, susceptible of supination and pronation—we find a more close resemblance to those of the human skeleton. The ulna, the superior extremity of which always projects beyond that of the radius, has a shaft which gradually narrows from above downwards. Its inferior extremity is terminated by a round head in those animals in which the ulna is fully developed; in others, as it is atrophied, it ends in a thin, long process.

The ulna presents at its superior extremity a posterior process, the olecranon, which forms the point of the elbow. We find on the anterior surface of the same, another process, the coronoid.

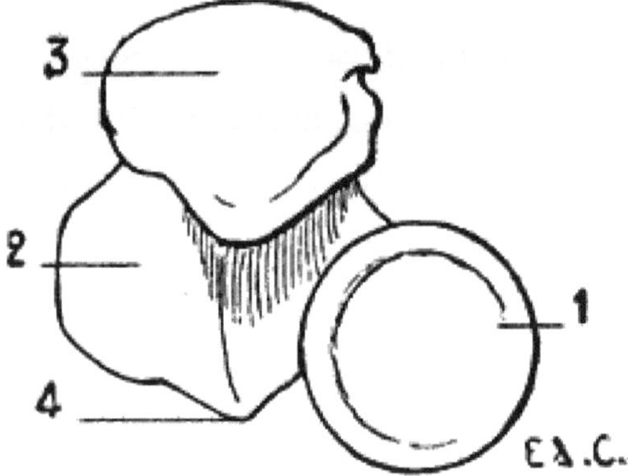

Fig. 25.—Superior Extremity of the Bones of the Human Forearm: Left Side, Superior Surface.

1, Radius; 2, ulna; 3, olecranon process; 4, coronoid process.

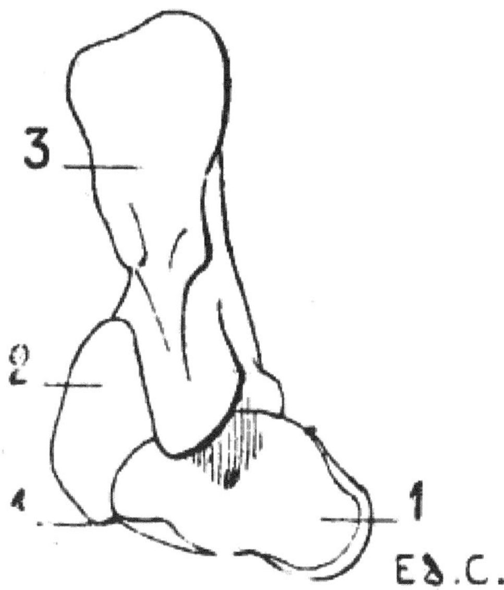

Fig. 26.—Superior Extremity of the Bones of the Forearm of the Dog: Left Limb, Superior Surface.

1, Radius; 2, ulna; 3, olecranon process; 4, coronoid process.

It is necessary to dwell on the relations of these parts. In man the head of the radius is situated at the anterior part of the external surface of the superior extremity of the ulna (Fig. 25); indeed, the small sigmoid cavity with which the head articulates is situated on the outer side of the coronoid process, and this apophysis is placed in front. In the plantigrades and digitigrades the head of the radius is placed still more forward, so much so that it is situated almost in front of the superior extremity of the ulna (Fig. 26). In the unguligrades it is placed directly in front of this latter (Fig. 27).

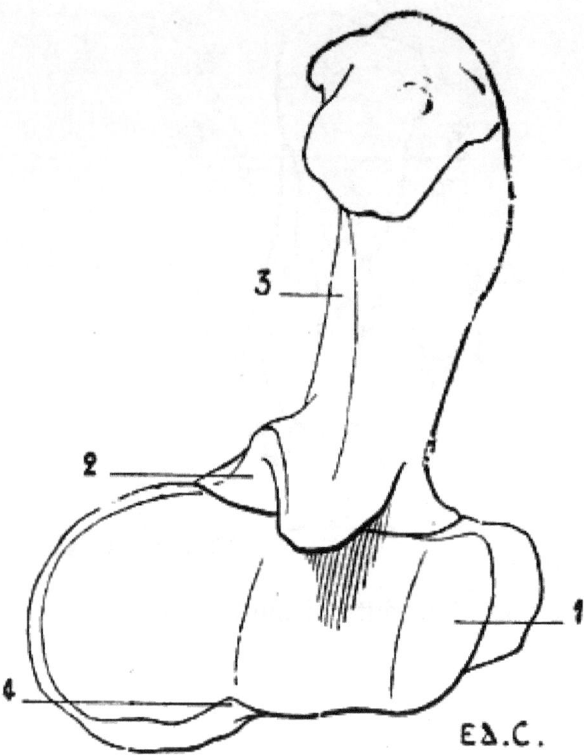

Fig. 27.—Superior Extremity of the Bones of the Forearm of the Horse: Left Limb, Superior Surface.

1, Radius; 2, ulna; 3, olecranon process; 4, coronoid process.

Further, the displacement of the radius is made at the expense of the superior extremity of the neighbouring bone; the radius appears to appropriate more and more the parts which in man belong exclusively to the ulna—for example, the coronoid process. In the plantigrades and the digitigrades half of the process still belongs to the ulna and the remainder to the radius. In the ungulates—the horse, for example—the coronoid process belongs to the radius; the ulna, situated behind the latter, is correspondingly diminished in size.

In brief, when we study this region of the skeleton in plantigrades, then in digitigrades, and finally in unguligrades, we find a kind of progressive absorption of one of the two bones (ulna) by the other (radius), which thus becomes the more developed.

It is easy to explain this partial disappearance of the ulna. When the forearm is capable of performing the movements of pronation and supination, the ulna is completely developed, for it is in its small sigmoid cavity that the head of the radius revolves, and it is around its inferior extremity, the head, that the corresponding extremity of the radius turns. But when the movements of rotation of the forearm do not exist, the inferior extremity of the ulna becomes functionally useless and disappears. As to its rôle in the movements of the region of the wrist, that is nil, for we may remember—we will observe it again when we come to treat of the articulations—that the hand articulates with the radius alone (radio-carpal articulation); this is the reason that, when the forearm possesses the fullest mobility, the hand follows the movements which that bone makes around the ulna.

It is not so with the articulation at the elbow-joint; there it is the ulna, which, with the humerus, forms the essential parts (humero-ulnar articulation); its olecranon process limits the movement of extension of the forearm. It is for this reason that, even in those quadrupeds in which the ulna is atrophied, the olecranon process presents a relatively considerable degree of development.

We know that on the posterior surface of the inferior extremity of the bones of the human forearm are grooves in which pass the tendons of the posterior and external muscles which, belonging to this region, are directed for insertion towards the hand.

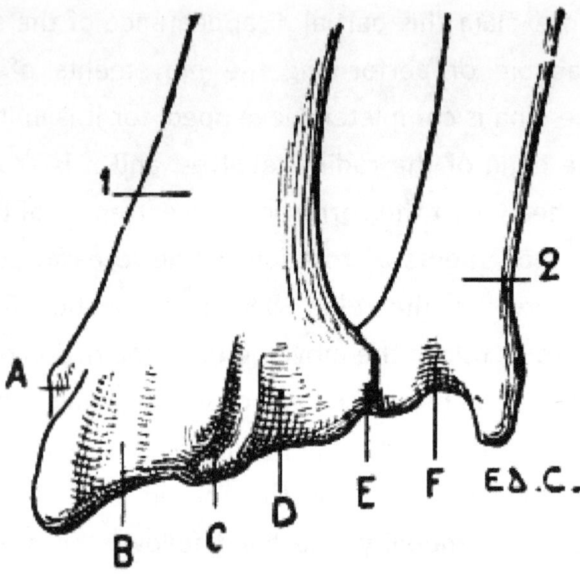

Fig. 28.—Inferior Extremity of the Bones of the Forearm of a Man: Left Side, Posterior Surface, Position of Supination.

1, Radius; 2, ulna; A, groove for the long abductor and short extensor muscles of the thumb; B, groove for the radial muscles; C, groove for the long extensor of the thumb; D, groove for the special extensor of the index finger and of the common extensor of the fingers; E, groove for the proper extensor of the little finger; F, groove for the posterior ulna.

In animals, because of the movement of rotation of the radius, the surface of this bone, which is anterior, corresponds to the posterior surface of the same in man. (To possess a clear conception of this, it is necessary to remember that, in this latter, the bones of the forearm are always described as in the position of supination; they are thus represented in Fig. 28. The direction of the surfaces of the radius is the reverse of that in animals, since the latter have the radius always in a state of pronation.)

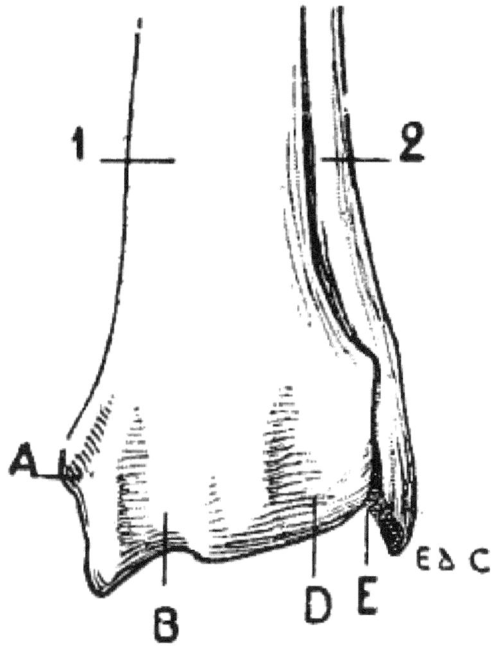

Fig. 29.—Inferior Extremity of the Bones of the Forearm of the Dog: Left Side, Anterior Surface, Normal Position—that is, the Position of Pronation.

1, Radius; 2, ulna; A, groove for the long abductor and for the short extensor of the thumb; B, groove for the radials; D, groove for the long extensor of the thumb, the special extensor of the index-finger, and the common extensor of the fingers; E, groove for the special extensor of the little finger.

Consequently it is on the anterior surface of the bone that we find the grooves concerning which it is necessary to give some details. Regarding them in passing from the radius towards the ulna, those grooves give passage to the tendons of the muscles whose names occupy the columns. The letters which are referred to each serve to define their order, and to facilitate reference to Figs. 28, 29, and 30.

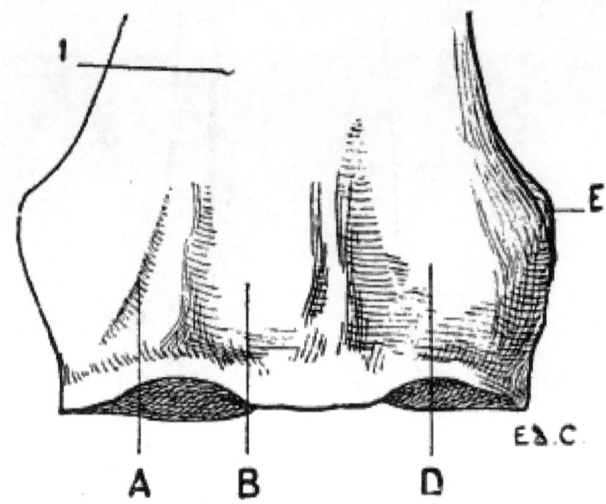

Fig. 30.—Inferior Extremity of the Bone of the Forearm of the Horse: Left Side, Anterior Surface.

1, Radius; A, groove for the long abductor and the short extensor of the thumb; B, groove for the radials; D, groove for the common extensor of the digits; E, groove for the special extensor of the little finger.

We should mention that the groove E is situated, both in man and in the dog, at the level of the inferior radio-ulnar articulation; but that in the horse, as the ulna does not exist at that level, the groove is situated on the external surface of the inferior extremity of the radius. It is necessary to add that, in some horses, the ulna is, nevertheless, represented in this region by a tongue-like process of bone; and in such cases the groove is situated in front of this process, at the level of the line of coalescence, which there represents the articulation.

Man.	Dog.	Horse.
A. Long abductor of the thumb, and short extensor of the thumb.	A. Long abductor of the thumb and short extensor of the thumb united as one muscle, *the oblique extensor of the metacarpus.*[11]	A. *Oblique extensor of the metacarpus*, the homologue of the long abductor of the thumb and the short extensor of the thumb, united as one muscle.
B. First and second external radials (*extensor carpi radialis longior* and *brevior*).	B. The two radials blended superiorly, distinct inferiorly; this is *the anterior extensor of the metacarpus.*	B. The radials represented by a single muscle, *the anterior extensor of the metacarpus.*
C. Long extensor of the thumb.	C. Long extensor of the thumb and special extensor of the index finger united superiorly. These muscles pass in the following groove.	C. The long extensor of the thumb and the special extensor of the index are absent.
D. Special extensor of the index finger and the common extensor of the fingers.	D. Common extensor of the digits and the two preceding muscles.	D. *Anterior extensor of the phalanges*, the homologue of the common extensor of the digits.
E. Special extensor of the little finger.	E. *Extensor of the third, fourth, and fifth digits*, or *the lateral extensor of the digits*, the homologue of the special extensor of the little finger.	E. Lateral extensor of the phalanges, the homologue of the special extensor of the little finger.
The posterior ulnar (*extensor carpi ulnaris*).	There does not exist on the forearm a groove for the posterior ulnar muscle, or *external flexor of the metacarpus.*	

[11] The words printed in italics are the names used in veterinary anatomy.

It is also useful to note, with reference to the groove F, in which passes, in man, the tendon of the posterior ulnar muscle, that, when the forearm is in pronation, the radius alone being displaced, we can only see this groove on the surface which looks backwards; and that it is then separated from the groove which contains the tendon of the special extensor of the little finger by an interval equal to the thickness of the head of the ulna.

[12] When the forearm is supinated, the two grooves are found, on the other hand, one beside the other: and the tendons which they contain are very naturally in contact.

[12] Édouard Cuyer, 'Shape of the Region of the Wrist in Supination and Pronation' (*Bulletin de la Société d'Anthropologie*, Paris, 1888).

In birds the forearm is flexed on the arm, and the latter being directed downwards and backwards, the former is, consequently, directed upwards and forwards. Further, because of the position of the humerus, which, as we mentioned, has its inferior extremity so turned that the surface which is anterior in man becomes external, the radius, instead of being outside the ulna, is placed above it. This latter is larger than the radius, but its olecranon process is very slightly developed.

The Hand

The hand in animals, as in man, is formed of three parts—the carpus, metacarpus, and fingers. In man, the forearm and the hand being described in the position of supination; the bones of the carpus are named in passing from the most external to the most internal—that is to say, from that which corresponds to the radial side of the forearm to that which corresponds to the ulnar side. In animals in which, as we know, but it is not unprofitable to repeat, the hand is in pronation, the radial side of the forearm being placed inside, we enumerate the carpal bones in counting the most internal as the first; this is the only method which permits us, in taking our point of departure from the human skeleton as our standard, to recognise the homologies of the bones of the carpal region.

These bones, eight in number, are arranged in two transverse rows, of which one, the first, is superior or anti-brachial; the other, the second, is inferior or metacarpal. Each of these rows contains four bones. Considered in the order we have indicated above—that is to say, proceeding from the radial to the ulnar side—they are thus named: scaphoid, semilunar, cuneiform, and pisiform, in the first row; trapezium, trapezoid, os magnum, and unciform, in the second. The number of these bones is not the same in all animals on account of the coalescence or absence of some. In each row the bones are placed side by side, with the exception of the pisiform, which being placed on the palmar surface of the cuneiform, produces a small projection in man, but a very pronounced one in quadrupeds.

The pisiform is called the *hooked bone* in some veterinary anatomies. If we consider the hook which it forms, we may recognise that the name is appropriate; but from the point of view of comparison with the human carpus, the name is unfortunate, for it creates confusion between the true pisiform (the fourth bone in the upper row), and the last bone in the lower row, which is the veritable unciform bone. We do not here seek for similarity of form, but homology of regions; and it is only by using the same names to denote the same things that we can succeed in determining such homology.

Taken as a whole, the bones of the carpus form a mass which, by its superior border, articulates with the bones of the forearm, and by its inferior border is in relation with the metacarpal region. Its dorsal surface (anterior in quadrupeds) is slightly convex; its palmar surface (posterior in quadrupeds) is excavated, and forms a groove in which pass the tendons of the flexors of the fingers. This last, in man, has the appearance of a gutter, because of the prominences caused by the projection of the internal and external bones beyond their fellows.

In quadrupeds the palmar groove is especially determined by the pisiform bone, of which we have just mentioned the great development.

The region occupied by the carpus, in the unguligrades, is known as the *knee*; it would have been more appropriately named had it been called the *wrist*.

The number of the metacarpal bones in mammals never exceeds five, but it often falls below it; the same is true for the digits. The first are generally equal in number to the latter; an exception is met with in ruminants, whose two metacarpals coalescing soon after birth, form but one bone; this, the *canon* bone, articulates with two digits.

The number of metacarpals and digits diminishes in proportion as the limbs cease to be organs of prehension, and become more exclusively organs of support and locomotion.

The number of phalanges is two for the thumb and three for each of the other digits; except in the cetaceans, in which they are more numerous.

In the bat, the metacarpals and phalanges are very long, and form the skeleton of the wing; these phalanges are not furnished with nails; the thumb, which is very short, is alone provided with one (Fig. 8).

With regard to the relative dimensions of the bones of the metacarpus, it is necessary to remember that, in the human being, the second metacarpal is the longest; then, in the order of decrease, come the third, fourth, fifth, and first. In quadrupeds we shall also find differences in length (see the chapter relating to the anterior limbs in certain animals), but the order of decrease is not always that which we have just mentioned.

In man the articular surface, situated at the inferior extremity of each of the metacarpals, is rounded, and is called the head. This allows the first phalanx, which is in relation with that surface, to be displaced in every direction; indeed, this phalanx can not only be flexed and extended, but it can also be moved laterally; this latter movement allows of the fingers being separated and drawn together.

In quadrupeds which can only perform the movements of flexion and extension of the digits—for example, the horse—the inferior extremity of the metacarpal has not a rounded head of a regular outline; it is marked by a prominent median crest, directed from before backwards, so that the articular surfaces, which fit more exactly, form a sort of hinge which allows of backward and forward movements only, and permits no lateral displacement. In man, at the level of the inferior extremity of the first metacarpal, in the vicinity of the articulation of this bone with the first phalanx of the thumb, we find two sesamoid bones—small bones developed in the fibrous tissue which surrounds the articulation. We also meet with such structures, but more rarely, at the level of the corresponding articulation of the index and auricular digits; and, more rarely still, at those of the middle and ring fingers. In quadrupeds, these bones are normally developed, and we shall see afterwards that in some animals, as they reach a considerable size, they are able to influence the external outlines; we shall see this, for example, in the horse.

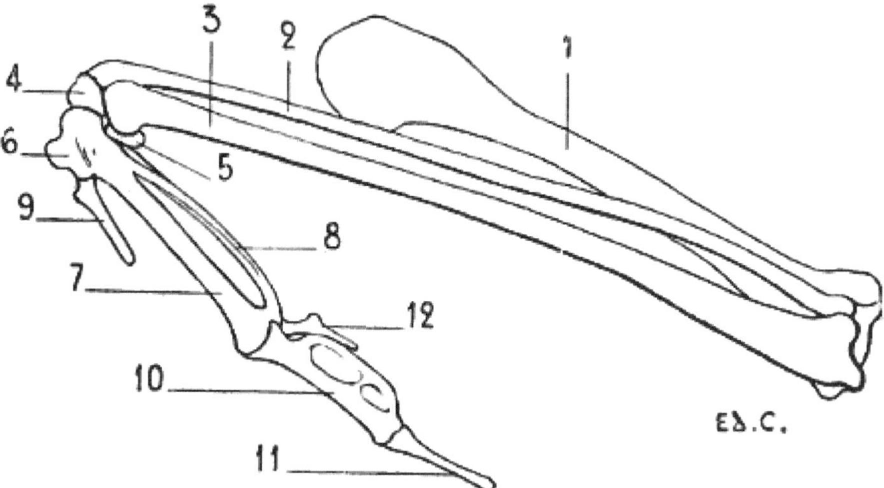

Fig. 31.—Skeleton of the Superior Limb of a Bird (Vulture): Left Side, External Surface.

1, Humerus; 2, radius; 3, ulna; 4, radial bone of the carpus; 5, ulnar bone of the carpus; 6, first metacarpal; 7, second metacarpal; 8, third metacarpal; 9, first digit, the homologue of the thumb; 10, first phalanx of the second digit; 11, second phalanx of the second digit; 12, third digit.

The hand, in birds, is directed obliquely downwards and backwards (Fig. 31). For the better understanding of its position in relation to the forearm, we should remember that this latter, as we have described, directed obliquely upwards and forwards, has the radius placed above the ulna; the hand being oblique in the opposite direction and placed under the forearm is, by this arrangement, inclined towards the ulnar border of the latter.

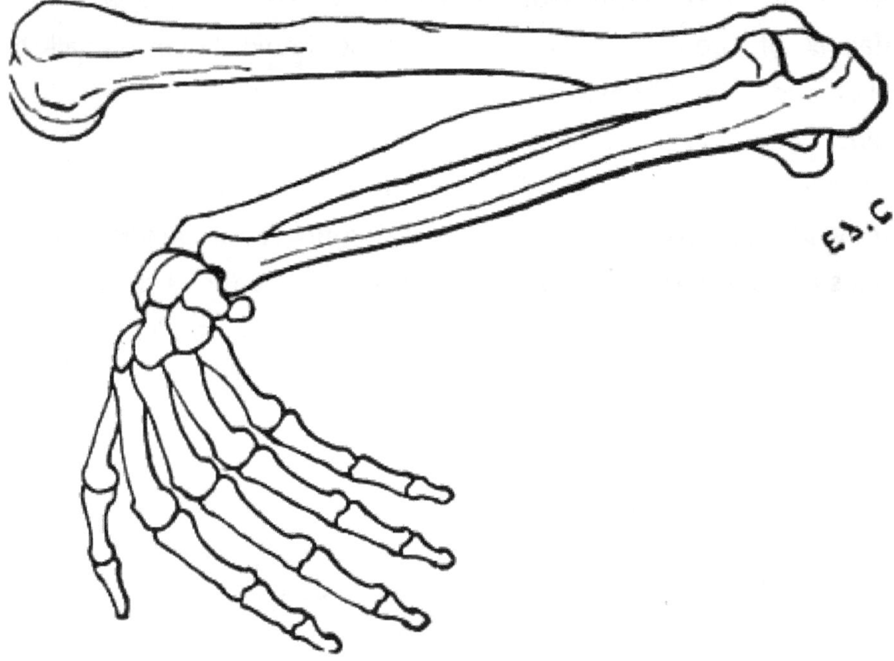

Fig. 32.—Superior Limb of the Human Being, the Different Segments being placed in the Attitude which the Corresponding Parts occupy in Birds: Left Side, External Surface.

For the rest, in order to be able to distinguish readily the corresponding parts in the hand of a bird and that of a man, we merely have to place the human forearm obliquely, in a direction upwards and forwards (Fig. 32), the radius being above; this position we can obtain by semi-pronation; then, to incline strongly the hand downwards and backwards, moving the ulnar border of the hand towards the ulna; the thumb is then anterior, the little finger posterior, and the palm of the hand is turned towards the trunk.

The carpus in birds is formed by two bones only, with which the skeleton of the forearm articulates. That which is in contact with the radius is called the *radial bone of the carpus*; and that with which the ulna articulates is named the *ulnar bone*.

The metacarpus is formed of three bones; the first, which is very short, is fused at its superior extremity with the adjoining one; this latter and the third, both longer than the first, but of unequal size, are fused at their extremities. The metacarpal, which articulates with the radial bone of the carpus, is larger than the one which is in line with the ulna. To the metacarpus succeed three digits, of which the central is the longest, and is formed of two phalanges; the other two are formed each by a small, stylet-shaped bone. The middle finger, situated on the prolongation of the metacarpal, which articulates with the radial bone of the carpus, has its first phalanx large and flattened transversely; this phalanx seems to have been formed by the union of two bones of unequal development; the second phalanx is styloid in form. As to the other two fingers, they are placed, one in front and the other behind; the first, which articulates with the short metacarpal, fused at its upper end with the principal bone of the metacarpus, in position represents the thumb. The other, which is the third finger, articulates with the inferior extremity of the thinnest bone of the metacarpus; it is sometimes closely united to the corresponding border of the first phalanx of the large—that is to say, of the median—digit.

The Anterior Limbs in Certain Animals

Plantigrades: Bear (Fig. 33).—The scapula of the bear approaches in shape to a trapezium, of which the angles have been rounded off. The anterior border (cervical) is strongly convex in the part next the glenoid cavity. The junction of the superior (spinal) and the cervical border forms almost a right angle, the summit of which corresponds to the origin of the spine. At its posterior angle there is a prominence, directed downwards, the surface of which is hollowed and is separated from the infraspinous fossa by a crest, so that at this level a third fossa is added to the infraspinous one. The neck of the scapula is but slightly marked. The acromion is prominent, and projects a little beyond the glenoid cavity.

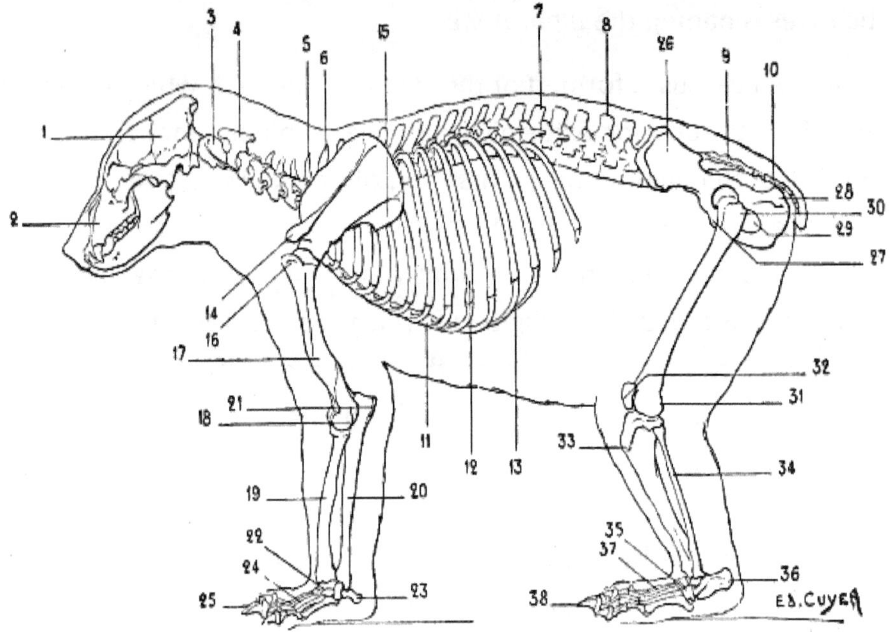

Fig. 33.—Skeleton of the Bear: Left Lateral Surface.

1. Cranium; 2, face; 3, atlas; 4, axis; 5, seventh cervical vertebra; 6, first dorsal vertebra; 7, fourteenth and last dorsal vertebra; 8, lumbar vertebræ; 9, sacrum; 10, coccygeal vertebræ; 11, sternum; 12, ninth and last sternal rib; 13, costal cartilages; 14, acromion process; 15, third fossa on the external surface of the scapula; 16, great tuberosity of the humerus; 17, musculo-spiral groove; 18, epicondyle; 19, radius; 20, ulna; 21, olecranon process; 22, carpus; 23, pisiform; 24, metacarpus; 25, phalanges; 26, ilium, external fossa; 27, pubis; 28, tuberosity of the ischium; 29, obturator foramen; 30, great trochanter of the femur; 31, condyles of the femur; 32, patella, or knee-cap; 33, anterior tuberosity of the tibia; 34, fibula; 35, tarsus; 36, calcaneum, or heel-bone; 37, metatarsus; 38, phalanges.

The clavicle is rudimentary, but, as an example of the complete development of this bone in plantigrade quadrupeds, we may cite the marmoset.

The humerus is furnished at its superior extremity with a large tuberosity, wide, and situated in front of the head of the bone; the effect of this is that the bicipital groove is internal. As in man, the great tuberosity does not reach so high as the humeral head, but it approaches more nearly to that level. The deltoid impression is very extensive, and descends pretty far down on the body of the bone. The epitrochlea is prominent; the epicondyle is surmounted by a well-marked crest, curved and flexuous in outline.

The articular surface, which is in contact with the radius, is not a regularly formed condyle; it is a little flattened on its anterior surface, and presents at this level a slight depression which corresponds to a small eminence on the anterior aspect of the superior extremity of the radius. The surface which articulates with the ulna, viewed on its anterior aspect, has the shape of a slightly-marked trochlea; except at the level of the internal lip, which, as in man, descends lower than the surface for articulation with the radius (condyle). Behind, the trochlea is more clearly defined.

The bear possesses a considerable power of rotation of the radius; the bones of the forearm are joined only at their extremities, while in the remainder of their extent they are widely separated. The ulna terminates below in a head and a styloid process; these articulate with the two last bones of the first row of the carpus—viz., the cuneiform and pisiform. The bones of the carpus are seven in number, the scaphoid and the semilunar being fused together.

The metacarpals, five in number, differ very little from one another in regard to length, though they increase in size from the first to the fifth; this may be demonstrated by looking at the palmar surface of the hand. It is the reverse of that which we find in man, for the fifth metacarpal is the thickest of all, and the first is the most slender.

At the level of each metacarpo-phalangeal articulation are two sesamoid bones.

The third digit is the longest. The terminal phalanges present two very different portions: one, the anterior, is curved and pointed; it serves to support the nail, whose shape it assumes; the other, posterior, forms a sort of sheath into which the base of the nail is received.

The inferior portion of the posterior surface of this latter part articulates with the second phalanx in the case of each of the last four digits, but with the first phalanx in the case of the thumb.

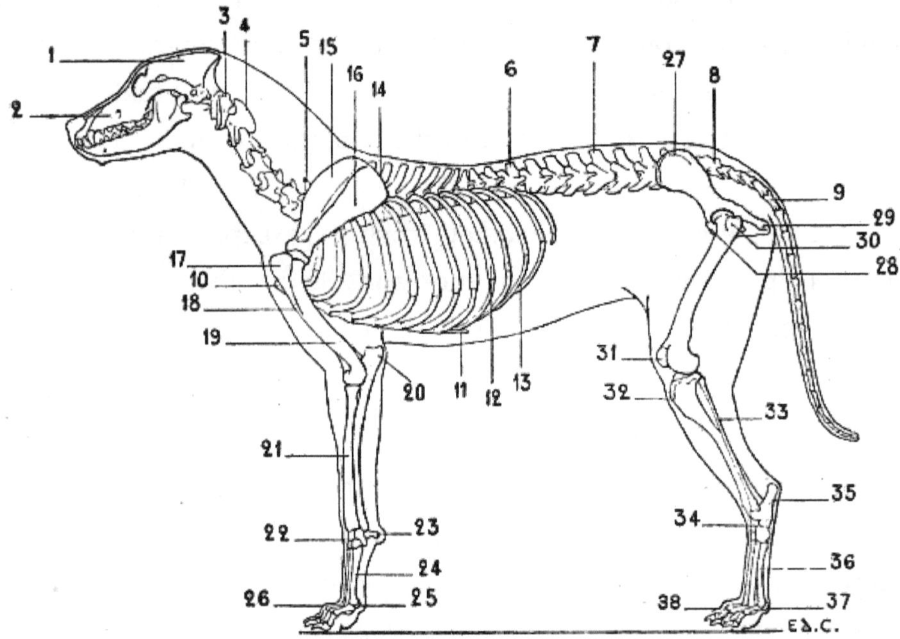

Fig. 34.—Skeleton of the Dog: Left Lateral Surface.

1, Cranium; 2, face; 3, atlas; 4, axis; 5, seventh cervical vertebra; 6, thirteenth and last dorsal vertebra; 7, lumbar vertebræ; 8, sacrum; 9, coccygeal vertebræ; 10, anterior extremity of the sternum; 11, xiphoid appendix; 12, ninth and last sternal rib; 13, costal cartilages; 14, spinal border of the scapula; 15, supraspinous fossa of the scapula; 16, infraspinous fossa of the scapula; 17, great tuberosity of the humerus; 18, deltoid impression; 19, musculo-spiral groove; 20, olecranon process; 21, radius; 22, carpus; 23, pisiform; 24, metacarpus; 25, sesamoid bones; 26, phalanges; 27, ilium, iliac crest; 28, pubis; 29, tuberosity of the ischium; 30, great trochanter of the femur; 31, patella, or knee-cap; 32, anterior tuberosity of the tibia; 33, fibula; 34, tarsus; 35, calcaneum, or heel-bone; 36, metatarsus; 37, sesamoid bones; 38, phalanges.

Digitigrades: Cat, Dog (Fig. 34).—In these animals the anterior (cervical) border of the scapula is convex; the posterior (axillary) border is straight or slightly concave. The supraspinous and infraspinous fossæ are of equal extent (Figs. 35 and 36). The neck is short. The spine of the scapula becomes more and more prominent towards its inferior extremity, where it ends in a twisted and inflexed portion, which represents the acromion process; this process terminates at the level of the glenoid cavity. The coracoid process is represented by a small tubercle, slightly curved inwards; this tubercle is situated above the glenoid cavity, at the inferior part of the cervical border.

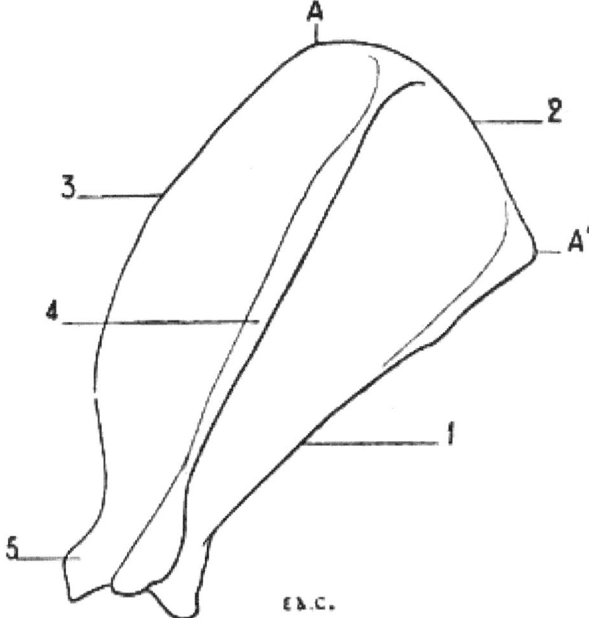

Fig. 35.—Scapula of the Dog: Left Side, External Surface.

1, Posterior or axillary border; 2, superior or spinal border; 3, anterior or cervical border; 4, spine of scapula; 5, coracoid process; AA′, length of spinal border.

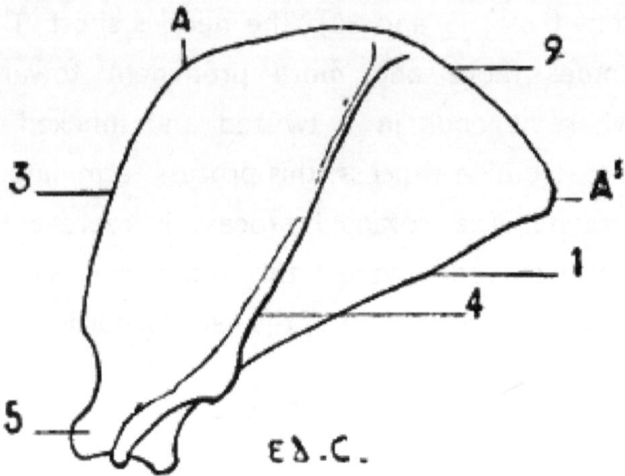

Fig. 36.—Left Scapula of the Cat: External Surface.

1, Posterior or axillary border; 2, superior or spinal border; 3, anterior or cervical border; 4, spine of the scapula; 5, coracoid process; AA', length of the spinal border.

In the dog, the posterior angle, formed by the junction of the axillary and the superior (spinal) borders, is obtuse; the spine rises perpendicularly from the surface of the bone. The width of the scapula, measured at the level of the spinal border (from A to A', Fig. 35), equals about half the length of the spine. We must, however, make an exception for the turnspit dog, in which the superior border equals three-fourths of that length. The scapula is, in this case, of a more compact type; it is broader, but shorter. In the cat, the anterior outline of the scapula, formed by the union of the cervical border and the corresponding half of the spinal, is more convex; the posterior angle is not obtuse, as in the dog. The spine is bent slightly downwards and backwards; before terminating in the acromion process it presents a triangular projection, the apex of which is directed downwards. The tubercle which represents the coracoid process is curved inwards more strongly than that of the dog, thus resembling more closely the appearance of this process in the human being.

All proportions considered, the scapula of the cat is broader than that of the dog; its width, measured along the length of its spinal border (from A to A', Fig. 36), equals three-fourths of the length of the spine.

The clavicle is rudimentary; it is, however, better developed in the cat than in the dog. The clavicle of the cat is represented by a small, elongated bone, curved in outline, the convexity being turned forward; it is united to the acromion and the sternum by ligamentous fibres; that of the dog is merely a scale-like osseous plate situated on the posterior surface of a muscle of this region (see Figs. 16 and 17).

The humerus is long and twisted in the shape of an S. The inferior articular surface has the form of a simple pulley, for the condyle is very slightly marked. The internal part of this articular surface descends lower than the external; this condition resembles that found in the human being, where the inner lip of the trochlea is lower than the condyle.

In the dog, the olecranon fossa communicates with the coronoid by an opening.

In the cat, there is a supra-epitrochlear canal (see Fig. 19), but no olecranon perforation.

The bones of the forearm articulate at their extremities. The body of the radius is united to the body of the ulna by a short, thick, interosseous ligament; the fibres of this ligament, though short, do not prevent the production of some movements at the articulations of the bones.

The radius so crosses the ulna that above, it is in front and external to the latter, while below, it is internal. This bone is flattened from front to back, and slightly convex anteriorly. Its superior extremity is formed, externally, of a portion which represents the head of the radius in man; internally, by another portion which represents half of the coronoid process of the ulna, which, in the human being, belongs exclusively to the latter (the encroachment of the radius on the ulna). This extremity is surrounded with a vertical articular surface which is placed in contact with a small cavity which is hollowed out on the ulna (the lesser sigmoid cavity); and presents at its superior aspect a surface which articulates with the inferior extremity of the humerus. The shaft of the bone has on its internal border rugosities analogous to the imprint of the pronator radii teres of the human skeleton; these rugosities, indeed, give insertion to a muscle of the same function, and bearing the same name. The inferior extremity, broader than the superior, is hollowed on its external aspect by a small cavity which receives the inferior extremity of the ulna; its inferior surface (concave) articulates with the carpus; its anterior surface (the homologue of the posterior surface of the corresponding extremity of the human radius) presents grooves which serve for the passage of the tendons of the muscles which pass from the forearm to the back of the hand. (For the names of the muscles whose tendons pass in these grooves, see Fig. 29.)

The ulna is furnished at its superior extremity with an olecranon process, which is more prominent than that of the human ulna; this process is compressed laterally, and its internal surface is hollowed; there we also find a great sigmoid cavity, and a coronoid process situated at the internal part of the anterior surface, a process which, as we have previously shown, it shares with the radius.

The shaft of the bone, prismatic and triangular, diminishes in thickness as it approaches the lower extremity, which articulates with the corresponding extremity of the radius. In the dog, the ulna terminates inferiorly in a blunt point, without enlargement, analogous to the head of the human ulna; in the cat, by a head which is prolonged into a styloid process, by which it articulates with a portion of the carpus.

The carpus consists of seven bones—three in the superior row and four in the inferior. In the superior row the scaphoid and semilunar bones are fused together. The pisiform is elongated and expanded at its two extremities; it forms a prominence which, directed backwards, projects beyond the level of the other bones of this region.

The metacarpal bones are five in number; they are enumerated from within outwards; they articulate with the carpus and with each other. The inferior extremity of each metacarpal bone presents the form of a condyle in front; and is divided behind so as to form two lateral condyles, which are separated by a median crest; on these posterior condyles are applied two small sesamoid bones. The metacarpal bone of the thumb is very short; the third and fourth are the longest. The metacarpus, as a whole, is directed vertically.

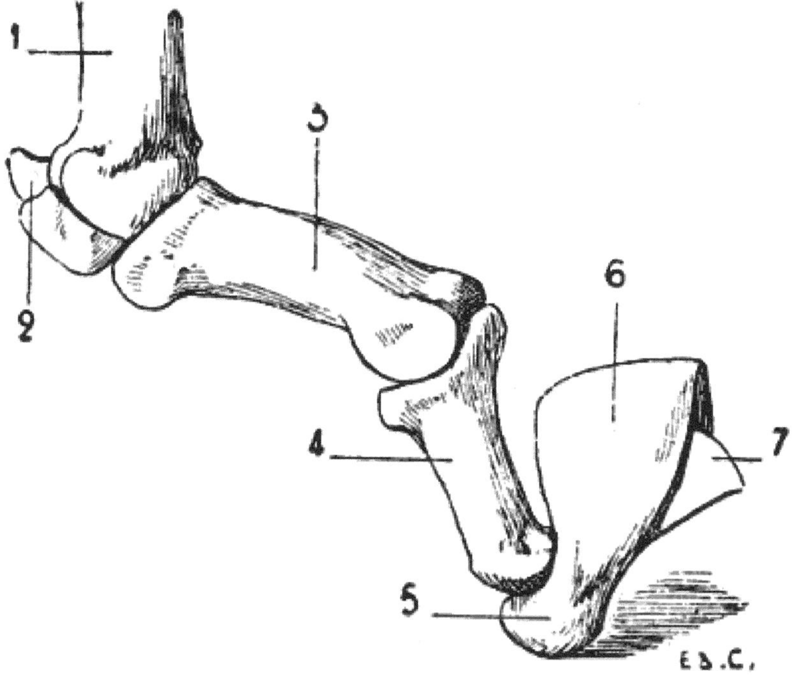

Fig. 37.—Skeleton of the Finger of a Felide (Lion): Left Side, Internal Surface.

1, Metacarpus; 2, sesamoid bones; 3, first phalanx; 4, second phalanx; 5, third phalanx; 6, gutter for the reception of the base of the nail; 7, prominent osseous crest formed to lodge in the concavity of the nail.

The phalanges are three in number for each finger, except the thumb, which has but two. The first phalanx, directed almost horizontally forwards, is the longest; the second is directed downwards and forwards; the third consists of two portions: a posterior part, which forms a sort of sheath into which the base of the nail is received; and an anterior, conical in form, and curved in crochet shape, which forms a support for the nail (Fig. 37).

The third and fourth digits are the longest; the second and fifth are of equal length; the thumb is the shortest; it does not touch the ground, and does not even reach the articulation of the metacarpal bone and first phalanx of the second finger.

In the cat, the metacarpal bone of the thumb, although shorter than any of the others, is quite as thick. The third digit is a little longer than either the second or fourth. In animals of this genus, the claws, in the condition of repose, are retracted, and removed from the ground; this prevents their being worn, and thus preserves their sharpness. At such times the third phalanx is received into a groove which is found on the external surface of the second phalanx. In the dog, the claws are not tractile.

[58]

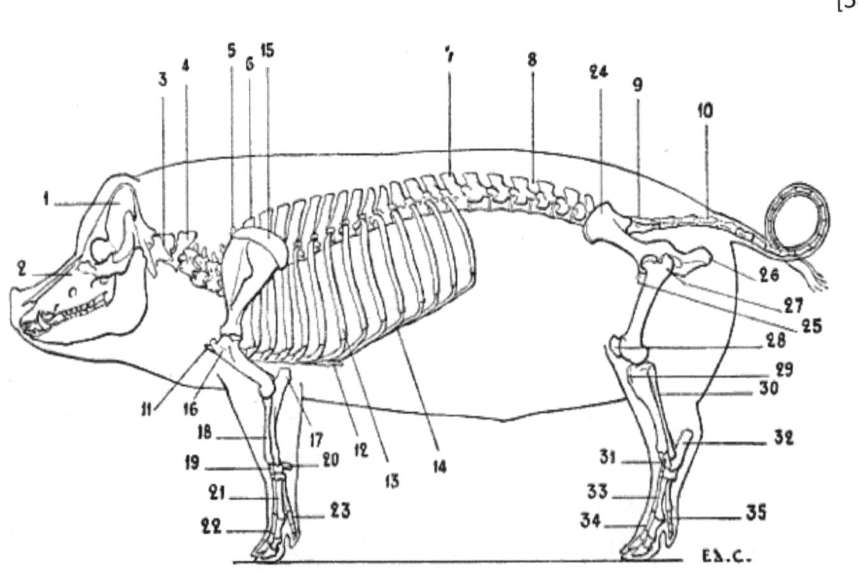

Fig. 38.—Skeleton of the Pig: Left Lateral Surface.

1, Cranium; 2, face; 3, atlas; 4, axis; 5, seventh cervical vertebra; 6, first dorsal vertebra; 7, fourteenth and last dorsal vertebra; 8, lumbar vertebræ; 9, sacrum; 10, coccygeal vertebræ; 11, anterior extremity of the sternum; 12, xiphoid appendix; 13, seventh and last sternal rib; 14, costal cartilage; 15, cartilage of prolongation of the scapula; 16, great tuberosity of the humerus; 17, olecranon process; 18, radius; 19, ulna; 20, pisiform; 21, metacarpus; 22, phalanges of the two great toes; 23, phalanges of the external toe; 24, ilium; 25, pubis; 26, tuberosity of the ischium; 27, great trochanter; 28, knee-cap; 29, anterior tuberosity of the tibia; 30, fibula; 31, tarsus; 32, calcaneum; 33, metatarsus; 34, phalanges of the two great toes; 35, phalanges of the external toe.

Unguligrades: **Pig** (Fig. 38).—The scapula is markedly narrowed in the region above the glenoid cavity. The spine is atrophied at both its extremities, so that at its inferior part we do not find the acromion process. In its middle portion the spine is prominent, and presents a triangular process which turns backwards, overlapping a part of the infraspinous fossa; this latter is much larger than the supraspinous. The spinal border is surmounted by the cartilage of prolongation, the superior margin of which is convex; this cartilage extends posteriorly beyond the posterior (axillary) border of the bone.

The small tuberosity of the superior extremity of the humerus is but slightly developed; the great tuberosity, on the contrary, is very large. The bicipital groove is situated internal to this. The deltoid impression is scarcely marked.

The forearm is short, directed obliquely downwards and inwards, thus forming with the hand an angle, of which the apex is directed inwards. The two bones of the forearm are strongly bound to one another by an interosseous ligament, which is formed of very short fibres. The radius appropriates, at its superior extremity, the coronoid process of the ulna. The latter is, notwithstanding, well developed in the rest of its extent; it has a flattened shaft which almost completely overlaps the posterior surface of the radius; its inferior extremity reaches to the carpus.

The carpus is formed of eight bones—four in the superior row, and four in the inferior. The third bone of the superior row (cuneiform) is more in contact with the ulna than with the radius.

There are but four metacarpal bones; there is no metacarpal of the thumb. The two median metacarpal bones are the longest; they are those which correspond to the digits which alone touch the ground. The internal digit and the external one are thin and short; they are functionless, as a rule, taking no part in supporting the limbs on the ground. Notwithstanding this, they are formed, as the other digits, of a number of phalanges, which give them the semblance of perfect digits.

(We shall soon see that in certain animals there exist digits which, being incomplete with regard to the numbers of their constituent bones, more accurately merit the name of imperfect digits.)

The third phalanges are each enclosed in a horny hoof, to which the name of *onglon* has been given.

We have already drawn attention to the smaller lateral digits, and noted the general fact that they do not come in contact with the ground. It is necessary to modify this statement by adding that under certain conditions they give a slight amount of support; for example, when the individual is the subject of excessive obesity, the limbs yield under the weight, and the nails of the lateral digits may touch the ground.

A similar fact may be noticed in pigs of ordinary bulk at the moment when, during walking, each of the fore-limbs commences to bear the weight—that is to say, when it is directed obliquely downwards and forwards; then all the digits are in contact with the ground.

[61]

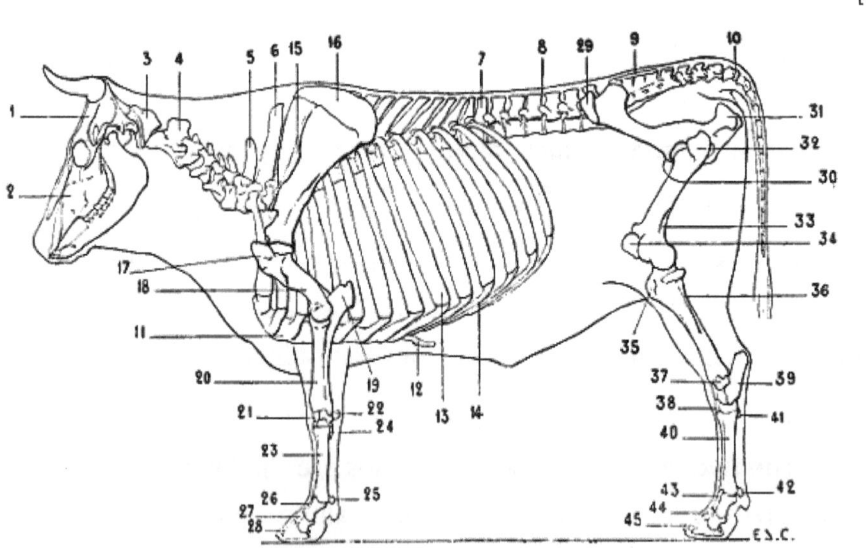

Fig. 39.—Skeleton of the Ox: Left Lateral Surface.

1, Cranium; 2, face; 3, atlas; 4, axis; 5, seventh cervical vertebra; 6, first dorsal vertebra; 7, thirteenth and last dorsal vertebra; 8, lumbar vertebræ; 9, sacrum; 10, coccygeal vertebræ; 11, sternum; 12, xiphoid appendix; 13, eighth and last sternal rib; 14, costal cartilages; 15, spine of scapula; 16, cartilage of prolongation of the scapula; 17, great tuberosity of the humerus; 18, musculo-spiral groove; 19, olecranon process; 20, radius; 21, carpus; 22, pisiform; 23, metacarpus; 24, rudimentary metacarpal; 25, sesamoid bones; 26, first phalanges; 27, second phalanges; 28, third phalanges; 29, anterior iliac spine; 30, pubis; 31, tuberosity of the ischium; 32, great trochanter; 33, supracondyloid fossa of the femur; 34, patella, or knee-cap; 35, anterior tuberosity of the tibia; 36, fibula; 37, coronoid tarsal bone; 38, tarsus; 39, calcaneum; 40, metatarsus; 41, rudimentary metatarsus; 42, sesamoid bones; 43, first phalanges; 44, second phalanges; 45, third phalanges.

Unguligrades (Ungulates): Sheep, Ox (Fig. 39).—The scapula, which is of elongated form, is very narrow in the vicinity of the glenoid cavity. The spine, which becomes more and more salient towards its inferior part, terminates abruptly in a border, which, forming an acute angle with the crest, produces a projection which represents the acromion process—a very rudimentary acromion, for it does not reach the level of the glenoid cavity. The supraspinous fossa is much smaller than the infraspinous; it hardly equals one-third the extent of the latter. The anterior border, thin and convex in its superior portion, is concave in the rest of its extent; the posterior border is thick and slightly concave; the spinal border is surmounted by the cartilage of prolongation. In the ox the spine of the scapula, in its middle portion, is flexed a little backwards on the infraspinous fossa.

The great tuberosity of the humerus is highly developed; its summit, very prominent, is flexed over the bicipital groove; a prominence of the small tuberosity also bends over the groove, with the result that at this level the latter is converted into a sort of canal. At the inferior extremity the condyle, although not large, is recognisable; for it is separated from the trochlea by a depression in form of a groove. In contrast to the condition found in man, the condyle descends to a level a little below that of the internal lip of the trochlea. (For the arrangement of the epicondyle and the epitrochlea.) In the sheep, the deltoid impression is but slightly marked; in the ox, it is more evident.

The forearm is directed obliquely downwards and inwards, so as to form, with the hand, an angle of which the apex is internal; this angular outline of the *knee* (wrist) is so characteristic of ruminants that the corresponding region of the horse, when salient inwards, receives the name of *ox-knee*. The radius bears the coronoid process, and the larger part of the articular surface which comes in contact with the inferior extremity of the humerus; the condyle and the trochlea articulate with the radius in front; while behind, the trochlea articulates with that part of the sigmoid cavity which belongs to the ulna. The posterior surface of the shaft of the radius is flattened; its anterior surface is slightly convex. The inferior extremity articulates with the carpus by a surface which is directed obliquely downwards and inwards. The shaft of the ulna is very slender, and fused in its middle third with the body of the radius; it terminates below, at the level of the external part of the inferior extremity of the radius, by a slightly expanded portion which, fused with this latter, forms the articular surface for the carpal bones.

In the ox the forearm is short; in the sheep it is proportionally longer.

The bones of the carpus are six in number—four in the upper row, and two in the lower; they form an irregular cuboid mass which contributes to the formation of the region known as *the knee* in ruminants, as in the horse; we have already remarked that the name 'wrist' would be more accurate. The anterior surface in its foremost part is vertical, and is slightly convex from side to side. At its posterior and external part the pisiform bone forms a prominence.

The metacarpus consists of two bones only—one, well developed, which is known as the principal metacarpal, or the *canon* bone (this is the name given to the region in the hoofed animals); and a rudimentary one, which is situated at the superior and external aspect of the preceding metacarpal. Sometimes there is found a third metacarpal at the internal aspect; but, when present, it is but very slightly developed.

The principal metacarpal consists of two metacarpals fused together; on this account the bone is longitudinally marked in the median line by a slight depression which marks the junction of the two bones of which it is formed. In some ruminants (certain species of chevrotains) the coalescence does not take place, and the two metacarpals remain separate.

The anterior surface of the principal metacarpal is convex transversely; its posterior surface is flattened. The superior extremity of this bone articulates by two facets with the two bones of the inferior row of the carpus; on the internal part of the anterior surface of this extremity is found a tubercle. The inferior extremity is divided into two parts by a fissure or notch; each part is articular, and consists of two separate condyles, which are separated from each other by an antero-posterior crest; on each side of this crest, and behind, are found two sesamoid bones. As for the external rudimentary metacarpal bone, it is nothing more than a small, short tongue of bone; which, in goats and sheep, is often absent.

The division of the inferior extremity of the principal metacarpal into two parts is correlated with the two perfect digits which give the foot of the ruminant its forked appearance. Each digit consists of three phalanges, which are directed obliquely downwards and forwards; further, these phalanges are inclined a little outwards from the axis of the limb, so that the two digits diverge from each other as they descend.

The first phalanx, which is the longest, articulates superiorly with the principal metacarpal; its inferior extremity terminates in a trochlea, and the lip of this, which is situated towards the axis of the limb, descends lower than that of the opposite side; this arrangement is correlated with the divergent direction of the digits. The second phalanx has its superior extremity moulded on the trochlea which terminates the extremity of the first; its inferior extremity is articular, and elongated from before backwards. On the posterior surface of this extremity is found a sesamoid bone.

With regard to the third phalanx, it presents the form of a triangular pyramid, and displays a postero-superior concave surface with which the second phalanx articulates; an anterior, convex surface, which terminates in a point on its anterior part; and an internal surface, which is flattened. The third phalanx of each digit is contained in a hoof (*onglon*).

There is also found in ruminants two imperfect rudimentary digits, which are represented by two small bones situated behind the articulation of the metacarpal and the digits which we have just been studying. These rudimentary digits are each enveloped in a layer of horn; they constitute the *spurs*. The two digits of the ruminants represent the third and fourth fingers of the human hand; the two lateral digits, greatly atrophied, are the homologues of the second and fifth fingers; the thumb is not present.

It is the same as regards the metacarpal bones, which form, by their union, the principal metacarpal; the external represents the fourth metacarpal, and the internal the third. It is to the latter that the tubercle, of which we have already made mention, belongs; and with the signification of which, because it gives attachment to a muscle, we shall concern ourselves in the section on myology (see Radial Muscles).

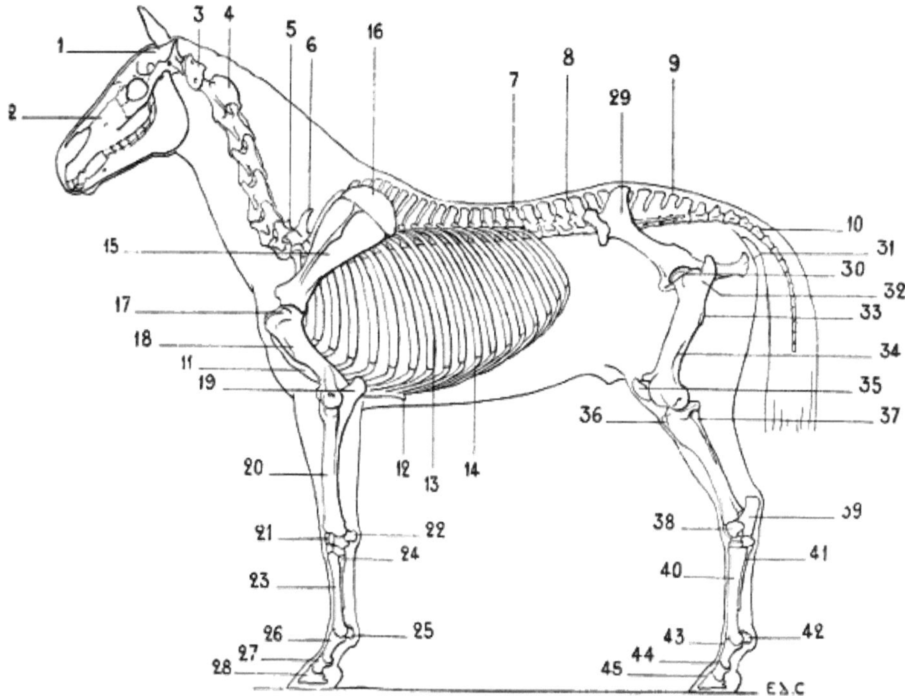

Fig. 40.—Skeleton of the Horse: Left Lateral Surface.

1, Cranium; 2, face; 3, atlas; 4, axis; 5, seventh cervical vertebra; 6, first dorsal vertebra; 7, eighteenth and last dorsal vertebra; 8, lumbar vertebræ; 9, sacrum; 10, coccygeal vertebræ; 11, sternum; 12, xiphoid appendix; 13, eighteenth and last sternal rib; 14, costal cartilage; 15, scapula; 16, cartilage of extension; 17, great tuberosity of the humerus; 18, deltoid crest; 19, olecranon process; 20, radius; 21, carpus; 22, pisiform; 23, principal metacarpal; 24, metacarpal, external rudimentary; 25, large sesamoids; 26, first phalanx; 27, second phalanx; 28, third phalanx; 29, ilium, showing external iliac fossa; 30, pubis; 31, tuberosity of the ischium; 32, great trochanter; 33, infratrochanteric crest, or third trochanter; 34, supracondyloid fossa of the femur; 35, knee-cap; 36, anterior tuberosity of the tibia; 37, the fibula; 38, tarsus astragalus; 39, calcaneum; 40, principal metatarsal; 41, rudimentary external metatarsal; 42, large sesamoids; 43, first phalanx; 44, second phalanx; 45, third phalanx.

Unguligrades: Horse (Fig. 40).—The scapula is narrow, compared with that of the animals we have just been considering. The anterior border is convex in its superior portion, and concave in its inferior; the posterior border is slightly hollowed out. The supraspinous fossa is less in extent than the infraspinous; but the difference is less than that between the same fossæ in the ox and the sheep; in the ox, as we have already indicated, the proportion is one-third; in the horse, one-half. The spine, which disappears at the extremities, is rough and thick in its middle third, there forming a kind of tuberosity—*tuberosity of the spine*. Above and in front of the glenoid cavity is found a strong process consisting of a rugous base, and a summit which is directed inwards. This forms a kind of hook curved towards the inside; it represents the coracoid process. The scapula is surmounted by the cartilage of prolongation, of which the superior border, which is thin and curved, is parallel to the superior border of the prominence of the withers; the cartilage forms, consequently, the lateral surface of this region. The cartilage of prolongation undergoes ossification in old horses. The humerus is short; the bicipital groove, situated on the anterior surface of the superior extremity, separates the greater tuberosity from the lesser, and is divided into two parts by a median ridge; it is this portion of the humerus which forms the prominence known as the *point of the shoulder*, or *point of the arm*. The deltoid impression well deserves the name of tuberosity which has been given to it, for it is very prominent; the musculo-spiral groove is very deep.

At the inferior extremity, the trochlea is large; the portion corresponding to the condyle of the humerus in man is, in proportion to the latter, of small extent. The olecranon fossa is deep. The epicondyle and the epitrochlea are somewhat different from those of the human bone. In the latter, the epitrochlea is salient towards the inner side, causing an increased transverse diameter of the inferior extremity of the humerus. In the horse—it is the same in ruminants—this tuberosity projects backwards, folds on itself in forming the internal boundary of the olecranon cavity, and exceeds in diameter, in the antero-posterior direction, the prominence of the epicondyle, which presents a nearly similar arrangement. This latter has, however, a part which, projecting externally, is situated at the inferior part of a crest, that forms the posterior boundary of the musculo-spiral groove. The result is that, contrary to the condition found in the human being, the epicondyle is more prominent transversely than the epitrochlea, but this latter is more salient on the posterior aspect. The epitrochlea and the epicondyle offer a larger surface for the origin of muscles of the forearm than the same prominences in the human bone do for the analogous muscles of the same region.

Some veterinary anatomists have given to the inferior and external articular surface of the humerus the name of trochlea; and to the internal one, that of condyle. On this account they designate the external prominence as the epitrochlea, and the internal one as the epicondyle. In addition to the fact that this point of view is not legitimate, it produces inevitable confusion when comparing the parts with those of the human humerus, and this confusion exists, not alone in describing the bone, but also in the description of the muscular attachments, and in the comparison of the muscles of the forearm of quadrupeds with the corresponding muscles in the human species.

The radius is placed in front of the ulna; its body, slightly convex forwards, has the anterior surface convex transversely, and the posterior surface plane in the same direction. It is to the external part of this latter that the ulna is applied, which is completely fused with the radius.

The superior extremity of the radius is a little larger than the inferior. Its superior aspect, concavo-convex, moulded on the inferior articular surface of the humerus, presents internally two cavities, which receive the lips of the trochlea, and, externally, another, smaller, cavity, which receives the condyle. The radius articulates with the trochlea and the condyle, having appropriated a portion of the ulna, as is proved by the presence of the coronoid process, which belongs to the former. This superior extremity presents, internally, a tuberosity into which the biceps is inserted; this is the bicipital tuberosity; and on the other side is another tuberosity, which is a little more prominent than the preceding.

The inferior extremity, which is flattened from before backwards, is furrowed on its anterior surface by grooves for the passage of muscles (the names of the muscles whose tendons pass in these grooves have already been given). It articulates at the lower end with the superior row of the carpus, and it terminates laterally in tuberosities: one, external, on which is found a groove for the tendon of the lateral extensor of the phalanges, the homologue of the special extensor of the little finger; the other, internal, is a little more prominent than the one we have just described. These tuberosities are visible under the skin which covers the superior and lateral parts of the region known as the *knee*; but which, we again repeat, is no other than the wrist.

The ulna has a triangular shaft, situated at the posterior surface of the radius, with which it is fused. It disappears completely at the level of the inferior third of the forearm. Occasionally, in some horses, the ulna is abnormally long, in the form of a slender tongue of bone; and extends to the neighbourhood of the external tuberosity of the inferior extremity of the radius (see Fig. 79, p. 196). Its superior extremity is chiefly represented by the olecranon process, which is voluminous in bulk, and forms the projection known as the point of the elbow. This process is flattened laterally; its internal surface is excavated; the anterior surface, which is concave, forms a part of the great sigmoid cavity; the remainder of the cavity is formed by the radius.

In the ass, the ulna is a little longer than in the horse—that is to say, it descends lower; and the radius is a little more convex anteriorly.

The carpal bones are seven in number—four in the superior row, and three in the inferior. The trapezium is wanting in the latter. Sometimes, however, in certain varieties of horses the trapezium is developed, but then it is no more than a very small osseous nodule. The pisiform bone, situated at the external part of the first row of bone, is prominent posteriorly. It is of rounder form and flattened from without inwards. It articulates with the trapezium and the radius. It presents, on its external surface, a groove for the passage of the tendon of the posterior ulnar muscle, which is named by veterinary anatomists the *external flexor of the metacarpus*.

The carpus, as a whole, is of an irregularly cuboid shape; its anterior surface, slightly convex from side to side, forms the skeleton of the region of the *knee* (wrist). The metacarpus is formed of three bones: the principal metacarpal and the two rudimentary ones.

The principal metacarpal, which forms the region of the *canon*, is directed vertically; its anterior surface is slightly convex transversely. This surface is covered by a number of tendons, which slightly alter its appearance; so that it is the principal base of this part of the fore-limb. Its posterior surface is flattened. The superior extremity of this metacarpal presents plane surfaces, variously inclined, with which the bones of the inferior row of the carpus articulate. On the anterior surface, and a little to the inner side, is found a tuberosity, which is destined for the insertion of *the anterior extensor of the metacarpus*, the homologue of the radial muscles. The inferior extremity is formed by two condyles, an internal and an external; between which is found a median crest.

This extremity, the superior extremity of the first phalanx, which articulates with it, together with two sesamoid bones—the great sesamoids—which are situated on its posterior surface, collectively form the region which from its rounded outlines is called the *ball*.

With regard to the rudimentary metacarpals, external and internal, to which some authors give the name of *fibulæ*, they are applied to the sides of the posterior surface of the principal metacarpal. They are elongated bones, of which the superior extremity, which is a little thickened, is called the *head*; the lateral bones of the second row of the carpus partly rest on the heads of these. They become more slender as they descend, and terminate opposite the inferior fourth of the principal metacarpal. Each ends in a slight swelling, to which the name *button* has been given. The internal one is the better developed.

The rudimentary metacarpals are vestiges of atrophied digits, as will be explained further on.

The single finger of the horse consists of three phalanges. The first phalanx, which is directed obliquely downwards and forwards, corresponds to the constricted region situated below the 'ball,' and known as the *pastern*. It is flattened from before backwards; its anterior surface is convex transversely, while the posterior surface is plane. Its superior extremity is moulded on the inferior extremity of the principal metacarpal, and its inferior extremity, which is smaller, presents a trochlea with which the second phalanx articulates. This is also directed downwards and forwards, and is shorter. It corresponds to the region which, situated between the pastern and the hoof, is known as the *cornet*.

The third phalanx, situated entirely within the hoof, has the same direction as the first and second. It is large and broad, and presents three surfaces separated by well-marked angular borders (see Fig. 96). The anterior surface is oblique downwards and forwards; it is convex transversely. The inferior surface is slightly hollowed, and is in relation with the sole, or plantar surface of the hoof.

The superior surface, which is articular, is divided by a median ridge into two lateral cavities, which correspond to the trochlea on the inferior surface of the lower extremity of the second phalanx. The inferior border corresponds in shape with the hoof. The superior border presents in its median part a projection, *the pyramidal eminence*, which prolongs at this level the anterior surface of the bone. Finally, the posterior border, which is concave, is in contact with a sesamoid bone, *the lesser sesamoid*, which increases the superior articular surface behind, and is also in contact with the second phalanx.

As we have just seen, the horse possesses but one digit. In the ancestors of the animal—that is, in the prehistoric species which are now extinct (*orohippus*, *miohippus*, *protohippus*, or *hipparion*)—the number of digits was larger; this fact conclusively proves that the rudimentary metacarpals of the existing horse are vestiges of digits which have disappeared through want of use. In the first of those ancestors—

orohippus—there were four digits; all save the first, the thumb, being then developed. In the others of the series there existed but three digits. It must, however, be noted that in those animals it is always the digit which corresponds to the middle finger of the pentedactyl hand that is longest. In other less ancient species the lateral fingers are reduced to the condition of mere splints of bone. It follows from what has been said that the digit which persists in the equine species should be considered as the third finger, and that the rudimentary metacarpals represent lateral digits considerably atrophied.

This disappearance of the lateral digits cannot excite surprise when we consider the functions of the organs. Becoming useless, they must undergo gradual atrophy from want of use.

There undoubtedly is, in this former existence of supplementary digits in the horse, something analogous to what we still find in the pig; where the two principal digits are accompanied by two shorter ones, which very probably, from their infrequent use, are destined to disappear in a more or less distant future.

Proportions of the Arm, the Forearm, and the Metacarpus

As a supplement to the study of the anterior limbs which we have just finished, it appears necessary to give some indications of the relative proportions of certain of the segments which form these limbs in the plantigrades, the digitigrades, and the ungulates.

First, we would remark that, in following this order of classification, the scapula becomes less and less narrow, and assumes a form more and more elongated. In order to convince ourselves of this, it will be sufficient to study the bone first in man, then in the bear, the cat, dog, ox, and finally in the horse.

As to the proportions of length, which are those we should chiefly study, we shall commence with the comparison of the forearm and arm—that is to say, the radius and the humerus. The radius is found to be longer in proportion to the humerus, as the number of digits is smaller, and the hand loses more and more the functions of an organ of prehension. In man, the radius is shorter than the humerus; in the horse, on the contrary, it is longer.

To give an idea of this proportion, we shall employ what is known as the antibrachial index. This index gives the relation which exists between the length of the forearm and that of the humerus; the length of this latter, whatever may be the actual measurement, is represented by a fixed figure, the number 100. A very simple arithmetical operation gives the proportion—

forearm × 100

humerus , the quotient obtained furnishes the index.

The index is less than 100 if the forearm is shorter than the bone of the arm. The index is more than 100 if, on the contrary, the forearm is longer.

In man, the radius is shorter than the humerus; indeed, in adult individuals of the white race the average index is 74.

In the bear, the length of the radius approaches closely to that of the humerus; the index is about 90. In the skeleton of a bear in the anatomical museum of the École des Beaux-Arts, the humerus is 33 centimetres in length, and the radius 30 centimetres.

In the cat, the radius is very little shorter than the humerus. In the dog they are equal. The antibrachial index of the latter is, accordingly, 100.

In the horse, the radius is longer than the humerus; the index is therefore above 100. Thus, in the skeleton of the horse which we have in the museum of the École des Beaux-Arts, the index is 113—length of humerus, 29 centimetres; length of radius, 33 centimetres. In other skeletons which we have measured we found: in one, 108—humerus, 34 centimetres; radius, 37 centimetres; in another, 116—humerus, 25 centimetres; radius, 29 centimetres.

The metacarpal bone undergoes, relatively to the humerus, a proportional elongation, analogous to that of the forearm.

In man, the length of the metacarpus is contained about 51⁄2 times in that of the humerus; in the bear, it is contained 4 times; in the dog, 21⁄2 times; in the horse, 11⁄3 times only.

It is well known that the proportions vary according to race, and that what we have here given are but the general indications.

The Articulations of the Anterior Limbs

The knowledge of human arthrology which we presume the reader to have previously acquired makes it unnecessary for us to enter into numerous details regarding the configuration of the articular osseous surfaces and the disposition of the fibrous bands that retain them in position. Accordingly, in the description which follows, and also in that of the articulations of the posterior limbs, we shall occupy ourselves but very briefly with the details above referred to, so as to devote ourselves especially to the indication of the movements—that is to say, of that which, while easily comprehended on recollection of former studies, presents the greatest interest from the artistic standpoint in these studies in comparative anatomy.

The Scapulo-Humeral Articulation.—The head of the humerus and the glenoid cavity of the scapula being in contact, the two bones are bound together by a rather loose articular capsule, which is strengthened by the muscles of this region which fulfil the function of active ligaments.

This articulation, so movable in every direction in the human species, is not so much so in quadrupeds; the arm in the latter, as also the shoulder, being kept in contact with the lateral region of the thorax by the numerous muscles which surround it.

Of the movements performed by the humerus, flexion and extension are the most extensive; those of abduction and adduction are much less so.

It is necessary, before proceeding further, to determine what the two principal movements which we have just mentioned really are, viz., flexion and extension.

We know that in man the displacements of the humerus which take place in the antero-posterior direction are known as movement or projection forwards, and movement or projection backwards, respectively. We do not say that the humerus is flexed or extended, because, in reality, on account of the position which the skeleton of the shoulder occupies, it is not able to flex or place itself on the line of prolongation of the scapula with which it articulates.

In quadrupeds it is not so. The humerus and the scapula are contained in almost the same vertical plane; and the bone of the arm can take, in relation to the latter, the positions characteristic of flexion and extension—that is, of approach to the scapula and removal from it.

What makes the meanings of these terms a little confusing is that, in human anatomy, some authors consider the backward movement of the humerus as extension, and the forward movement as flexion; in order to be able to compare these movements to those that the femur executes in relation to the pelvis.

Now, in our opinion, the indication of this correspondence is not absolutely necessary; since it ceases to be exact if we wished, from the point of view of the direction given to other segments of the skeleton, to establish the same relation between the elbow and the articulation of the knee.

It is therefore indispensable, when discussing quadrupeds, to discontinue these terms, in order the more readily to recognise that: in flexion the inferior extremity of the humerus is directed backwards; in extension, on the contrary, it is directed forwards. In the first case the humerus approaches the scapula; in the second, on the contrary, it moves away from it.

These movements, which take place during walking, are executed in the following manner: When one of the anterior limbs is at the end of that stage of progression which is called support (Displacements of the Limbs)—that is to say, during the time that the foot remains in contact with the ground, whilst the trunk is moving forward—the direction of this limb becomes more and more oblique downwards and backwards. At a certain moment the limb is raised from the ground, to be carried forwards, in order to be again pressed on the ground, and recommence a new resting stage. In these different phases the humerus is flexed. But at the moment that the limb, when carried forwards, is about to resume its contact with the ground it becomes directed obliquely downwards and forwards; then the humerus is in the position of extension.

During these movements of the humerus, there exists an essential factor—that is, the scapular balance. (It is the same as what occurs in man when he balances his arm in the antero-posterior plane.) When the humerus is flexed, the scapula moves in such a way that the superior portion projects forwards; when it is extended, the scapula, on the other hand, is inclined more backwards. But it is necessary to add that, during these displacements, the scapulo-humeral angle varies; it tends to close during the flexion of the humerus, and becomes more open during extension.

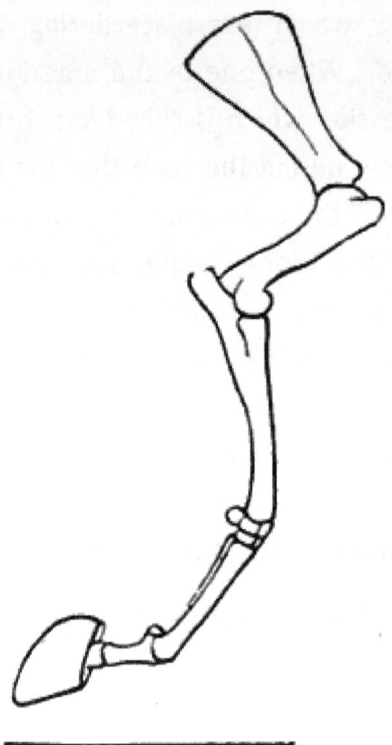

Fig. 41.—Flexion of the Humerus: Right Anterior Limb of the Horse, External Surface (after a Chromophotographic Study by Professor Marey).

The movements and the relations of the humerus and the scapula are clearly represented in Figs. 41 and 42, reproduced from the chromophotographic studies of Professor Marey—studies relative to the analyses of the movements of the horse.[13] They show clearly the movements of flexion and extension of the humerus, also the balancing of the scapula which accompanies the movements.

[13] E. J. Marey, 'Analyses of the Movements of the Horse by the Chromophotograph' (*La Nature*, June 11, 1898).

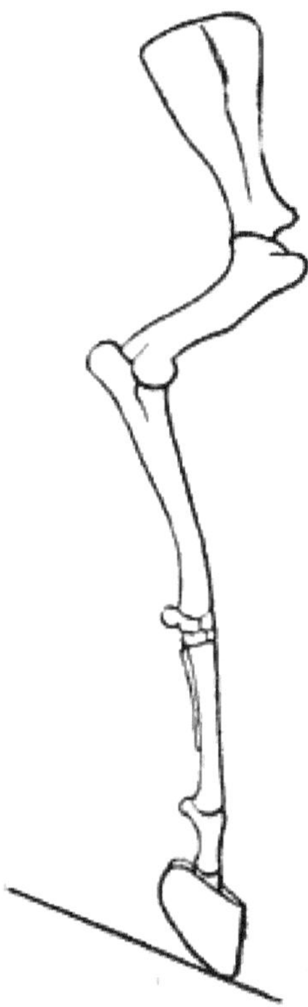

Fig. 42.—Extension of the Humerus: Right Anterior Limb of the Horse, External Surface (after a Chromophotographic Study by Professor Marey).

The Articulation of the Elbow, or the Humero-ulnar Articulation.—In this articulation, which is constructed in the form of a true hinge, the movements of flexion and extension alone are possible. In flexion, the forearm, directed forwards, is folded on the arm, with which, in certain circumstances, it comes in contact. For example, in a horse of mettle which leaps over an elevated obstacle, the animal forcibly raises his fore-limbs by flexing them. Flexion is produced to the same extent, and even more so, and for a longer period, in felides which crouch.

In extension, on the contrary, the forearm is carried backward. This movement being limited only by the contact of the tip of the olecranon with the bottom of the olecranon fossa of the humerus, the forearm is enabled, in this case, to move until it is in line with the arm. For example, during walking, when one of the anterior limbs, having reached the end of its resting stage, is considerably inclined downwards and backwards.

The apex of the olecranon process—that is to say, the point of the elbow—forms a marked prominence, more salient in flexion than in extension, as in the corresponding region of the human elbow.

The Radio-ulnar Articulation.—It is in the dog and the cat, in which the two bones of the forearm articulate by their extremities only, and remain separate in the rest of their extent, that the articulations call for special notice.

In the upper part, the radius rotates on itself; while below, it rotates around the ulna. It follows that the forearm, which in all quadrupeds is in a state of permanent pronation, can, in carnivora, take the position of supination, or rather, of demi-supination. In fact, whatever be the mobility of the two bones of the forearm, the movement is not able to bring the palmar surface to the front, but only to direct it towards the median line.

The Articulation of the Wrist.—Here are found, as in man, three superimposed articulations: the radio-carpal, intercarpal, and carpo-metacarpal.

If we remember the movements which take place at the plane of these articulations in man, and take account of the fact that the mobility of the limbs is reduced just in proportion as they are simplified in structure so as to become organs of support only, we can easily comprehend that, in the horse and the ox, and, in a word, animals that have a canon bone, the movements of the wrist are little varied in character, while in carnivoræ, on the other hand, they are relatively more so.

We will remember that in the ox and the horse the region of the wrist is called the *knee*.

In flexion, the hand is bent backwards; in extension it is carried forwards. These two movements take place especially in the radiocarpal and intercarpal articulations. In the first of these articulations, it is the superior row of the carpus which glides backwards and forwards on the corresponding articular surface of the forearm. In the second articulation, it is the second row which moves; gliding on the inferior articular surfaces of the row above it. This inferior row carries the metacarpus with it; for the carpo-metacarpal articulation is much less mobile than either of the other two.

In flexion, the articular surfaces are separated from one another in front; and the changes of form which result from this are noticeable on the anterior surface of the 'knee.' Moreover, at that moment this region contrasts markedly in its outlines with the parts above it and below it—that is to say, with the corresponding surfaces of the forearm and of the canon bone.

As for the lateral movements, by which the hand is inclined outwards and inwards in its movements at the wrist, they exist to an appreciable extent in the cat and the dog only; in order to understand this, it is enough to compare the shape of the articular surfaces of this region in carnivora and the horse, for example. In the latter, those surfaces are almost plane; in the cat, on the contrary, they are curved (inferior surface of the forearm, concave; superior border of the carpus, convex). These latter, then, are, in form, similar to those which exist at the same level in the human being; this explains the possibility of analogous movements of the whole hand—that is to say, of the movements of abduction and adduction.

The Metacarpo-phalangeal Articulations.—With regard to the mobility, it is in these articulations, as in those of the wrist—that is to say, although in all quadrupeds the first phalanges can be flexed and extended on the metacarpus, it is only in the cat and dog that lateral movement is possible. Indeed, in the horse, in which the principal metacarpal terminates inferiorly in two convex surfaces, which are separated by a crest; and where the whole articulates with a cavity on the superior extremity of the first phalanx; because of the hinging of these surfaces, there can only be movements of opening and closing of this articulation. The first phalanx is directed backwards during flexion and forwards during extension. In the dog and the cat the digits can be separated from each other, and also drawn together—that is to say, abducted and adducted; but, as in man, these movements can be made only when the first phalanges are in the state of extension. During flexion they are impossible, because of the tension of the lateral ligaments, which increases as the flexion is more pronounced. This can be demonstrated, for example, in the cat, which, in order to separate the digits, opens the hand widely—that is to say, forcibly raises the first phalanges.

The Interphalangeal Articulations.—The phalanges are in contact with one another by surfaces, which, on one side, are of trochlear form, and, on the other, are moulded on these trochleæ; accordingly, at the level of these articulations, the movements of flexion and extension only can take place.

In the felidæ, the claws which the third phalanges bear cannot be utilized when the latter are in a state of extension, at which time, being forcibly raised, they are, in fact, placed on the outer sides of the phalanges, which are grooved to receive them. But when the animal wishes to use them, it flexes those third phalanges, of which the terminal extremity is then projected forward, and the claws are ready to fulfil their function. But at the same time it extends the first phalanges, to produce a certain tension of the flexors of the digits, and thus enable the latter to act with greater efficacy, with a minimum of contraction. We can demonstrate this action experimentally on ourselves. It is enough to carry the first phalanges forcibly into a state of extension; the third phalanges then become flexed, quite spontaneously, by the tension of the tendons of the flexors which are inserted into them.

At the same time, if we examine the felidæ which we have taken as examples, when the first phalanges are in the state of extension, the digits will be found to be separable, as we have already indicated in connection with the metacarpo-phalangeal articulations, with the result that the claws are then able to lacerate a wider surface.

The extension of the ungual phalanx, which determines the retraction of the claw and stops its action, is the mechanical result of an elastic, fibrous apparatus which is attached to each of the third phalanges, and has its origin of the second.

THE POSTERIOR LIMBS[14]

The posterior limbs are divided, as are the inferior limbs of the human being, of which they are the homologues, into four parts: pelvis, thigh, leg, and foot.

[14] Examine Figs. 21, 33, 34, 38, 39, 49.

The Pelvis

The pelvis, which incompletely limits the abdominal cavity, inferiorly in the vertical position of the body and posteriorly in the normal attitude of quadrupeds, is formed by the iliac bones and sacrum—the coccyx forming a prolongation of the latter. We have already described the two latter in connection with the vertebral column, of which they form the inferior or posterior portion or segment, according to the attitude of the individual.

The Iliac Bone.—The iliac or coxal bone, is a paired or non-symmetrical bone, united below to its fellow of the opposite side, while it is separated from it above by the sacrum.

In all animals, as well as in man, the iliac bone, at the beginning of life, consists of three parts, which afterwards unite and fuse together and join at the middle of the bottom of a deep cavity which is situated on the outer aspect of the bone—the cotyloid cavity.

Of those three portions when examined in the human iliac bone, that above the cavity is the ilium; that on the inside is the pubis; and the last, the lower one, is the ischium. In quadrupeds, the iliac bone being, in its entirety, directed much more obliquely downwards and backwards, the relative position of these constituent parts is a little modified: the ilium is in front, the pubis is still internal, but in a more inferior position, and the ischium is behind the cotyloid cavity. We notice this peculiarity of the development of the iliac bone because it is customary to continue to apply to the osseous regions which correspond to these parts the names by which they were known when independent bones.

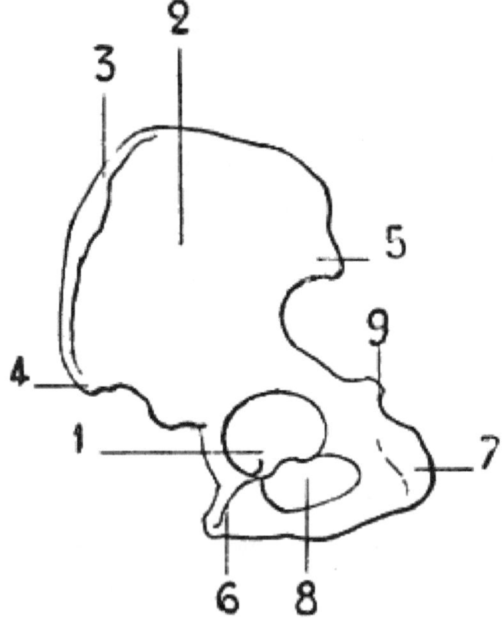

Fig. 43.—The Left Iliac Bone of the Human Being: External Surface, placed in the Position which it would occupy in the Skeleton Of a Quadruped.

1, Cotyloid cavity; 2, ilium; 3, iliac crest; 4, anterior iliac crest; 5, posterior iliac spine; 6, pubis; 7, tuberosity of the ischium; 8, obturator foramen; 9, ischiadic spine.

The bones which form the skeleton of the pelvis of quadrupeds are proportionally more elongated and less massive than those of the human pelvis (Figs. 43 and 44).

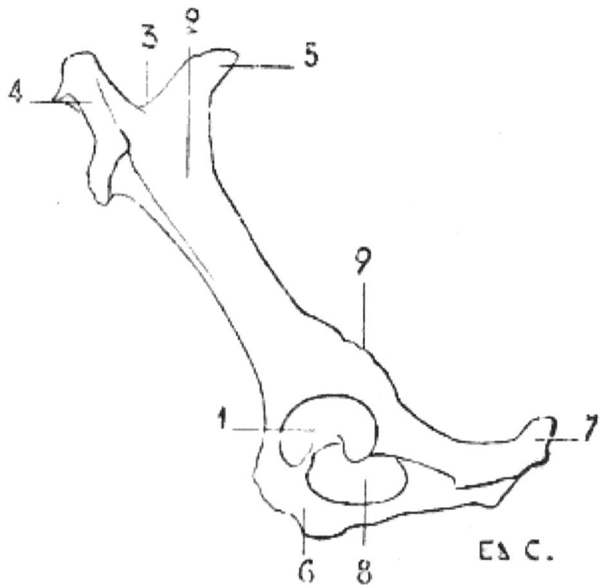

Fig. 44.—Left Iliac Bone of A Quadruped (Horse): External Surface.

1, Cotyloid cavity; 2, ilium, external iliac fossa (directed upward in the horse); 3, iliac crest; 4, anterior iliac spine (directed inwards in the horse, it is the angle of the haunch); 5, posterior iliac spine (directed inwards in the horse; it is the angle of the haunch); 6, pubis; 7, tuberosity of the ischium; 8, obturator foramen; 9, ischiadic spine, or subcotyloid foramen.

We find, on the external surface of the iliac bone, the cotyloid cavity, whose border is interrupted by the cotyloid notch; a deep notch which looks downwards.

In front of this cavity is the ilium. This portion, narrow in the part which is next the cavity, is directed forwards and upwards, expanding more and more as it passes upwards. It presents an external or superior surface (external in some animals, superior in others), which recalls the external iliac fossa; and an internal or inferior surface, at the superior part of which is found the auricular surface for articulation with the sacrum.

The anterior border of the ilium is rough; this is the iliac crest, at the extremities of which we find, below or outside, a prominence which corresponds to the anterior superior iliac spine of man; and internally another projection which corresponds to the posterior iliac spine.

Immediately above the cotyloid cavity is a rough crest, which is known as the *supracotyloid crest*, which is, however, no other than the homologue of the sciatic spine. In front of this prominence, the border of the ilium, which is notched, forms the great sciatic notch.

If, still taking the cotyloid cavity as the point of departure, we proceed inwards—that is, towards the median line of the body—we find the pubis; if in a posterior direction, the ischium. These two portions, pubis and ischium, limit an oval orifice, the subpubic foramen.

In the human skeleton, the pubis of one side is united to that of the opposite side, to form the pubic symphysis. In the animals which we are now studying a portion of the ischium enters into the formation of the symphysis; in other words, it is formed, not only by the body of the pubis, but also by the descending branch or ramus of the pubis and a portion of the ascending branch or ramus of the ischium, which are fused with those of the opposite side. It results that, though in the human being the symphysis is short and the ischio-pubic arch large, in quadrupeds it is the opposite. In them the arch is a mere slot, and being formed by the ischium alone, merits the name of the ischial arch. The ischio-pubic symphysis is very large, and forms a horizontal surface relatively extensive, a sort of floor, on which rest certain organs which occupy the cavity of the pelvis.

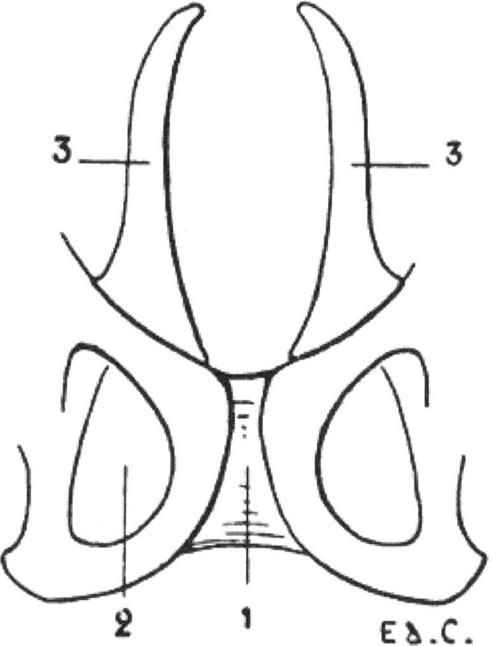

Fig. 45.—Pubic Region of the Pelvis of a Marsupial (Phalanger, Fox).

1, Symphysis pubis; 2, obturator foramen; 3, marsupial bones.

The posterior and external angle of the ischium is rough and prominent; it is the tuberosity of the ischium. This forms a projection under the skin; it also does in man when the trunk is strongly inclined forwards, while the thighs are maintained in the vertical position. In marsupials—opossum, kangaroo, and phalanger—the pelvis at its pubic region is surmounted by two bones, situated one on each side of the median line, and arranged in the form of a fork of two prongs (Fig. 45). These, which are called *marsupial bones*, support the pouch which, in animals of this genus, lodges their young, which, at the time of birth, are incapable of supporting a separate existence, their development being absolutely incomplete.

In the cetaceans—for example, the dolphin—because of the absence of posterior limbs, the pelvis is represented by two separate bones only, which have no connection with the vertebral column. In birds, the pelvis is remarkable for its elongated form (see for its form Fig. 21, and for details Fig. 46). The cotyloid cavity is pierced by an opening, and presents on its posterior border, which is here a little prominent, a surface with which the great trochanter is in contact.

The ilium is very highly developed, and is fused in the median line with the ilium of the opposite side, the last dorsal vertebræ, the lumbar vertebræ, and the sacrum. Because of these relations with the dorsal vertebræ, it is in contact anteriorly with the last ribs, which consequently emerge from each side of the iliac region of the pelvis.

The ischium forms a plate of bone which, in part, closes the external portion of the cavity of the pelvis. Its superior border is separate for a certain distance from the external border of the ilium; there is thus left an opening of more or less considerable size, which represents or takes the place of the great sciatic notch.

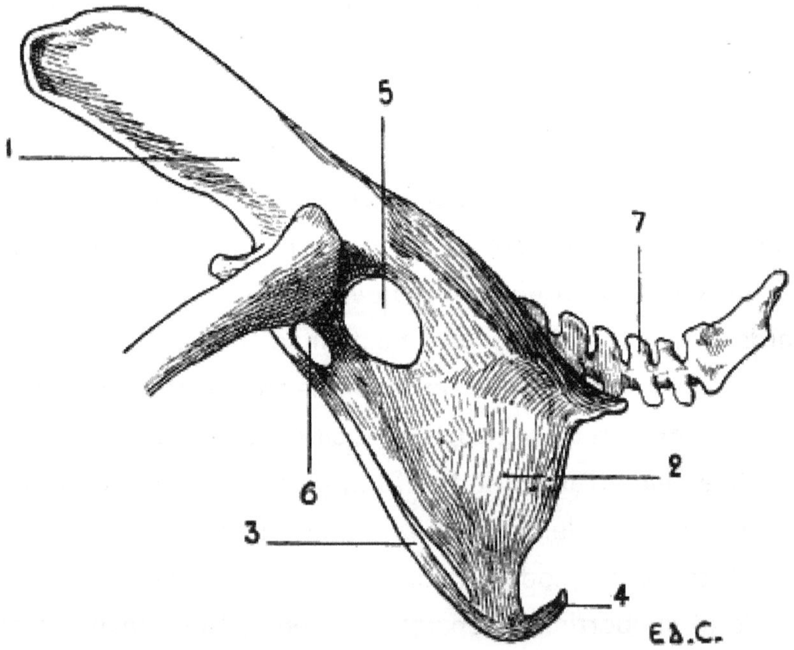

Fig. 46.—Pelvis of a Bird (the Cock): External Surface, Left Side.

1, Ilium; 2, ischium; 3, pubis; 4, inferior extremity of the pubis; 5, sciatic foramen; 6, oval foramen, homologous to the obturator; 7, coccygeal vertebræ.

The pubis, long and slender, is in connection with the inferior border of the ischium, of which it follows the general direction; and circumscribes with this latter, below the cotyloid cavity, an oval orifice, which is the homologue of the obturator foramen. Its inferior extremity reaches beyond the corresponding part of the ischium, bending towards the middle line, but without joining the pubis of the opposite side. On this account there is no symphysis pubis in birds. Nevertheless, an exception must be noted in the case of the ostrich, the pubic bones of which meet in the middle line, and are articulated in form of a symphysis.

The Thigh

A single bone, the femur, forms the skeleton of this portion of the lower limb.

The Femur.—The bone of the thigh is, in man, directed downwards and inwards; this obliquity, we may remind the reader, is due to the difference in length of the two condyles which form its inferior extremity; the internal is the more prominent, the result of which is that when the femur is held vertically, the internal condyle descends lower than the external. Now, as those two articular expansions rest on the horizontal plane formed by the upper extremity of the tibia, it follows that the superior part of the femur inclines towards the side of the shorter condyle—that is to say, outwards—and that, the leg being vertical, it and the bone of the thigh unite in forming an angle, of which the apex is directed towards the inner side of the knee.

In many mammals the two condyles are equally prominent, the result of which is that the femur inclines neither inwards nor outwards, but is contained in a plane parallel to the axis of the trunk; while the leg is included in the same plane. Nevertheless, although contained in the plane which we have just indicated, the femur is obliquely placed, and directed downwards and forwards; it accordingly forms, with the pelvis, an angle, of which the opening is directed to the anterior aspect of the body.

In reptiles and in birds the femur and leg are both placed in the same plane, but this plane is not parallel to the axis of the trunk. This is the result, on the one hand, of the thorax being wide, and, on the other hand, of the femur, which is directed forwards, being in contact by its anterior extremity with the lateral aspect of the costal region, it is thus necessarily placed in a direction forwards and outwards, and the knee is further removed from the axis of the trunk than is the articulation which unites the thigh with the pelvis.

The femur, like the humerus, is almost completely enveloped by muscular masses, which bind it to the lateral walls of the abdomen. Its inferior extremity alone is free, and is always the more so in proportion to its elongation—that is to say, as it belongs to an animal whose foot is more divided. The femur in this respect conforms to the law which we have indicated in connection with the bone of the arm, in which the development, as to length, is in proportion to the division of the hand.

If we compare the femur of certain animals with that of man, we see that the corresponding details of form are readily recognisable, but they are slightly modified. Thus, on examining the superior extremity, we find there a head, a neck, a great trochanter, and a lesser; but the neck is usually short and thick, and the great trochanter does not occupy the same level with regard to the articular head of the bone. In man, the great trochanter does not rise to the level of the head of the femur; in the dog and the cat it approaches that level; in the horse and in ruminants it rises above it.

With regard to the inferior extremity, its surfaces undergo modifications which are further accentuated as we pass from the digitigrades to the ungulates, or unguligrades. We know that in man the femoral trochlea is continuous behind, without interruption, with the condyles—that is to say, that each of the condyles is the continuation of one of the lips of the trochlea. We have just said that the trochlea is continuous without interruption with the condyles; this is accurate. Nevertheless, we must remark that, at the level of the junction of these surfaces, the bone presents a slight constriction, which is more marked on the external than on the internal aspect. This constriction, which is but slightly marked in man, is accentuated in the dog and the cat; in the ruminants and the solipeds it is still more pronounced so that we may say that in these latter the trochlea and the condyles are almost completely separated.

There is another modification in regard to the prominence and extent of the two lips of the trochlea. In man, the external lip of the trochlea reaches higher than the internal, and it is more prominent in front. In the dog, these lips are equal with regard to thickness, but the external still reaches higher than the internal; in the cat, they are equal in every respect; in ruminants and solipeds the internal lip is wider, thicker, and rises higher than the external.

In animals the trochlea is, as a general rule, narrower than in man, and the condyles are more prominent posteriorly; so that, when viewed from one of the lateral aspects, the inferior extremity of the femur is, in them, better developed in the antero-posterior direction.

In birds, the femur is shorter than the bones of the leg; its great trochanter is in contact with a prominence which occupies the posterior part of the border of the cotyloid cavity. Instead of articulating at the level of the knee, with the knee-cap and tibia only, as in man, it articulates, in addition, with the superior extremity of the fibula. A similar arrangement is found in marsupials and reptiles.

The Knee-cap.—This bone, developed in the thickness of the tendon of the triceps muscle of the thigh, is in contact, by its posterior surface, with the femoral trochlea. The two articular surfaces which are applied to the lips of the trochlea present, with regard to their extent, an inequality which is in proportion to the arrangement which we have above indicated—that is, while in man it is the external surface which is the larger, in the horse it is the internal. We shall see what the general form of the knee-cap is when we come, later on, to study more particularly the posterior limbs of some animals.

The Leg

The skeleton of the leg consists of two bones: the tibia and the fibula. The tibia is the more internal and the larger of the two; the fibula is slender, and situated on the outer side, and a little posterior to, the preceding. The fibula is more or less developed according to the species; in some it is complete, in others it is very much atrophied.

This peculiarity may be compared with that which we have drawn attention to regarding the development of the ulna; but here the seriation is less distinct. Not only in the different species, but even in the individuals of the same species, the development of the fibula presents little regularity. In quadrupeds, the bones of the leg are directed obliquely downwards and backwards, so that they form, with the femur, which is directed obliquely downwards and forwards, an angle, the apex of which is placed at the anterior surface of the knee.

Tibia.—The tibia of quadrupeds is readily comparable with that of man; as in the case of the latter, its shaft has three surfaces—an external, which is hollowed out in its upper portion, and becomes anterior below; an internal, slightly convex and subcutaneous; the posterior, which presents, in its superior part, a crest, the oblique line of the tibia, and some rugosities. The borders separate the surfaces. The anterior border, or crest of the tibia, is prominent in its superior part; below it gradually disappears in passing towards the internal aspect of the inferior extremity. The external and internal borders separate the corresponding surfaces from the posterior one.

The superior extremity is thick, and expands in forming three tuberosities: two lateral and an anterior. The anterior tuberosity, situated at the superior part of the crest of the tibia, is very prominent; for this reason the superior extremity is very much expanded in the antero-posterior direction—hence it results that this diameter is equal to the transverse, and sometimes even greater. In man, it is the latter which is the larger. The anterior tuberosity is visible under the skin.

The inferior extremity, less thick, is prolonged internally by a prominence which corresponds to the internal malleolus of man. In animals whose fibula is but slightly developed the tibia presents, on the external part of its inferior extremity, a small prominence, which replaces the fibular malleolus. The ruminants must, however, be excepted, in which we find in this region a special bone, which certain authors look on as the inferior part of the fibula. The inferior surface of this extremity of the tibia is articular; and is in contact with one of the tarsal bones, the astragalus. Because the superior surface of this latter has the form of a pulley, a pulley much more marked than that on the human astragalus, the corresponding surface of the tibia, which has the opposite form, presents two lateral cavities, separated by a median ridge, which is directed forwards and slightly outwards; this ridge projects into the groove of the pulley.

The Fibula.—This bone, situated at the back of the external surface of the tibia, is, as we have said, more or less developed. Its superior extremity, or head, articulates with the external tuberosity of the tibia. Its inferior extremity, when it exists—it is this which disappears in animals which have the fibula incompletely developed—forms a prominence which, placed on the external surface of the inferior extremity of the tibia, articulates with the astragalus, and recalls the external malleolus of man.

We have stated above that it is the inferior extremity of the fibula which disappears when the bone is incompletely developed; it is necessary to except the bat, in which the fibula, fairly well developed at its inferior extremity, by which it articulates with the tibia, thins off in its superior portion, and does not reach the corresponding extremity of the latter. Further, as in this animal the surface of the knee, which corresponds to the anterior surface of the same region in other animals, is turned backwards, the result is that the fibula is situated on the inner side of the tibia, instead of being placed on the outer.

The Foot

The foot, in animals, as well as in man, is formed of three portions, which, as we pass from the part which articulates with the leg towards the terminal extremity, are: the tarsus, the metatarsus, and the toes. These three portions are the homologues of the carpus, the metacarpus, and the fingers, which, as we have already seen in the case of the hand, are the osseous groups which form its skeleton. The tarsus is formed of short bones, as the carpus is; these are, in man, seven in number. The bones are arranged in two rows: one, the posterior, formed of two bones superimposed—the astragalus, by which the tarsus articulates with the leg, and the calcaneum, which forms the prominence of the heel; and an anterior row formed of five juxtaposed ones—the cuboid, situated externally, and the scaphoid internally, in front of which are found the three cuneiforms. To the tarsus succeeds the metatarsus, whose form reminds us very much of that of the metacarpals.

With regard to the toes, which we enumerate in proceeding from the most internal to the most external, they are formed of phalanges, which are three in number for the four outer toes; but the number is reduced to two in the case of the first—that is, the so-called great-toe.

The bones of the tarsus are not seven in all animals; they are fewer in ruminants and solipeds. We already know that, in the latter, the metacarpals and the digits are equally reduced in number; the same is the case for the metatarsals and the toes. We will analyze these differences when dealing with the species individually.

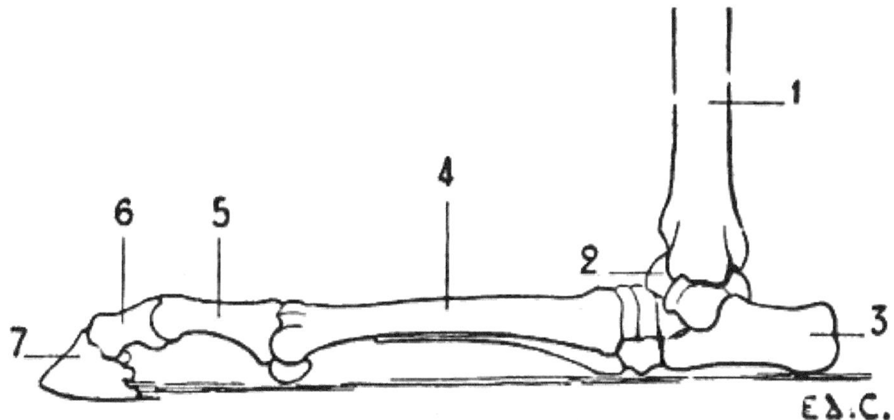

Fig. 47.—Posterior Limb of the Horse placed in the Position which it should occupy if the Animal were a Plantigrade: Left Limb, External Surface.

1, Tibia; 2, astragalus; 3, calcaneum; 4, metatarsus; 5, first phalanx; 6, second phalanx; 7, third phalanx.

When we studied the anterior limbs, we saw in passing from the plantigrades to the digitigrades, and finally the ungulates, or unguligrades, as the hand became hyperextended, the carpus was raised and more and more removed from the ground. We shall establish the existence of the same condition in the posterior limbs; in the plantigrades the tarsus rests on the ground; in the digitigrades it is removed from it; while in the unguligrades the distance which separates it from the point of support is still more considerable; and it is, indeed, necessary to imagine that if these latter were plantigrades, would occupy the position on the ground which is indicated by Fig. 47.

In veterinary anatomy the tarsus is called the *ham*; a name we adopt in conformity with usage, but which we cannot but regret, as in human anatomy the ham is the region of the posterior surface of the knee.

The general arrangement of the region of the digits of the posterior limbs in birds, presents some points of interest.

We shall merely say with regard to the metatarsus, that it is formed by a single bone, which in the cock is furnished towards its inferior third with a pointed process, the *spur*. At the inferior part, there is, however, found another, which is but very slightly developed, and with which the first phalanx of the innermost toe articulates.

The toes are, in the majority of species, four in number:[15] an internal, which is directed backwards, and corresponds to the great-toe; the others are directed forwards. This arrangement is constant in grallatores (wading birds), gallinaceæ[16] (domestic fowls), and raptores (birds of prey).

[15] In spite of the fact that the custom is to designate the terminal portions of the foot of birds by the name of digits, we prefer to employ here the terms *foot* and *toes*. In adopting this decision we believe we are acting according to a more didactic method. Homology of names should, in our opinion, always accompany homology of regions.

[16] With regard to the gallinaceæ, we must add that in certain varieties the number of toes is five; those which are directed forwards are three in number; the internal one which passes backward, is double. The two toes which are the subject of this special arrangement are placed very close together, and are nearly always superimposed. This condition is found in the Houdan and Dorking breeds.

In climbing birds (parrots, woodpeckers, and toucans), the innermost toe is not only directed backward, but the external toe accompanies it in that direction; consequently, there are two posterior and two anterior toes. Sometimes they are all directed forwards; this disposition is found in the martins. In some birds, the number of toes is reduced to three: the cassowary shows this reduction; in others, the number is still further diminished—the ostrich, for example, has but two.

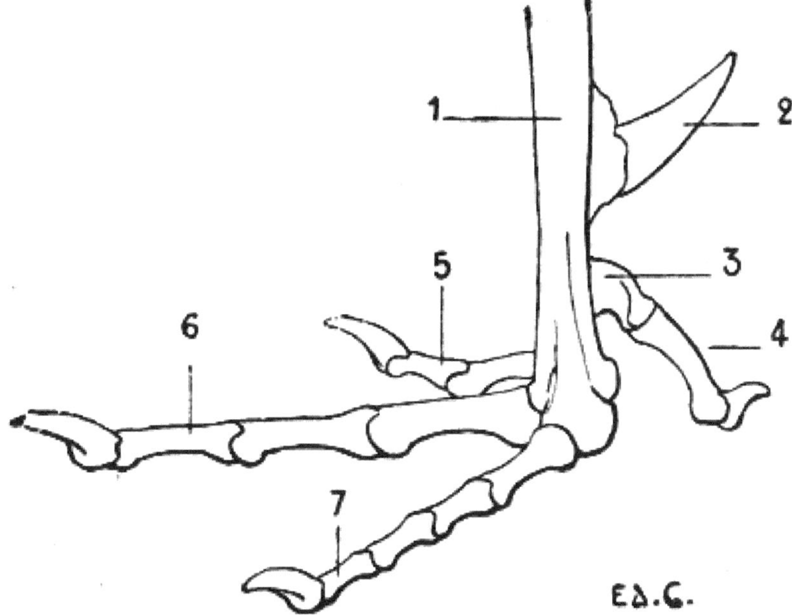

Fig. 48.—Skeleton of the Foot of a Bird (the Cock): Left Side, External Surface.

1, Metatarsus; 2, spur; 3, rudimentary metatarsal; 4, first toe; 5, second toe; 6, third toe; 7, fourth toe.

Further, we find that, in general, the number of the phalanges increases, when we examine the toes in commencing with the most internal (Fig. 48): this has two; then the following one three; that which comes next in order has four; and the most external toe has five. The phalanges of this last are short; so that, although it is formed by a larger number of bones, it is not the longest of the toes.

THE POSTERIOR LIMBS IN SOME ANIMALS.

Plantigrades: Bear (Fig. 33).—The external iliac fossa is very deep. The femur is longer than the bones of the leg; the great trochanter does not reach the level of the head of the femur. The fibula is well developed; it is united to the tibia at its superior and inferior extremities only.

The foot, which, as in the case of the hand, rests on the ground by the whole extent of its plantar surface, presents five toes; the shortest of these is the internal—that is, the toe which corresponds to the great-toe in man; the third and fourth are the longest, and they are almost equal; there is a very slight difference in favour of the fourth, which is slightly superior in dimensions to the third.

Digitigrades: Cat, Dog (Fig. 34).—The external iliac fossa, which looks outwards, is deep; the iliac crest is convex anteriorly, the convexity is continued from one iliac spine to the other.

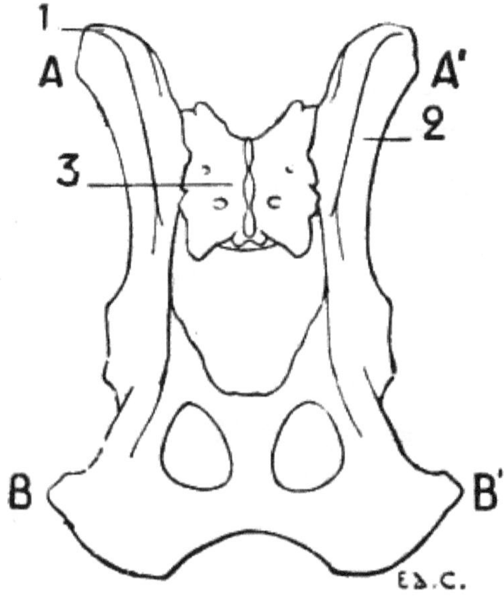

Fig. 49.—Pelvis of the Dog, seen from Above.

1, Iliac crest; 2, external iliac fossa; 3, sacrum; AA′, bi-iliac diameter; BB′, bi-ischial diameter.

In the dog, the distance which separates the anterior iliac spines is less than that which separates the ischia (Fig. 49). On a skeleton which we measured, the transverse diameter, the distance from the anterior iliac spine of one side to that of the opposite side, was 8 centimetres, whilst the distance which separated the ischia was 105 millimetres; on another skeleton, the first measurement was 127 metres, and the second was 146 millimetres. It seems to us unnecessary to multiply examples.

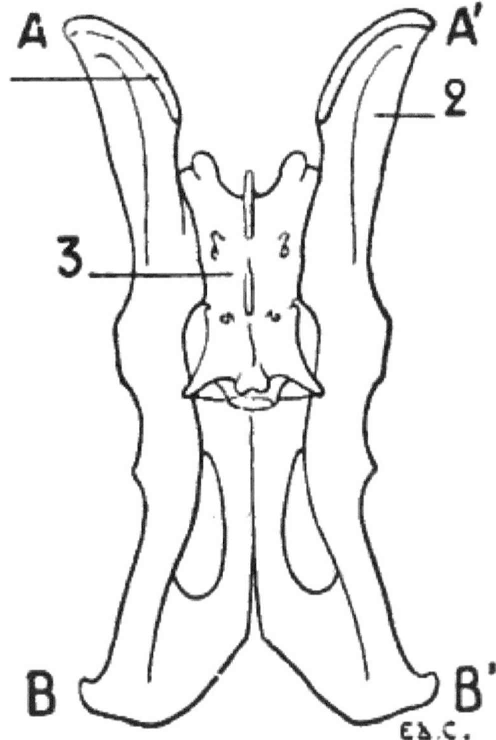

Fig. 50.—Pelvis of a Felide (Lion), viewed from Above.

1, Iliac crest; 2, external iliac fossa; 3, sacrum; AA′, bi-iliac diameter; BB′, bi-ischial diameter.

In the cat, the iliac spines are but slightly marked; the result is that the iliac crest is almost confounded with the inferior and superior borders of the ilium. The two diameters referred to above are almost equal (Fig. 50).

We draw particular attention to what we have just noted in regard to the transverse proportions of the iliac and ischiatic regions of the dog and the cat. These relations are evidently of importance with regard to shape, since the iliac crests and the ischia are noticeable beneath the skin.

In the dog, the shaft of the femur is slightly convex in front; but in the cat it is straight. The borders of the shaft are slightly marked, so that it is almost cylindrical. The *linea aspera*, less prominent than in man, gains in width what it loses in elevation; it constitutes what may almost be called a rough *surface*. This surface is narrower in its middle portion than at its extremities, where it bifurcates to go upwards to the two trochanters, and downwards to the two condyles. At the superior extremity, the neck is short, the great trochanter reaching almost to the level of the head of the femur; the digital *cavity*, which is situated on the internal surface of the great trochanter, is very deep. At its inferior extremity it projects strongly backward. The trochlea is narrow; in the cat its two lips are equally prominent, while in the dog the external is a little more elevated than the internal, which on its part is a little thicker. The trochlea is still more independent of the condyles than in the human femur; it is separated from these latter by a slight constriction.

The knee-cap is long and narrow.

The tibia of the dog is slightly curved from before backward: it has the form of an elongated S; this conformation is in great part due to the very marked projection of the anterior tuberosity and of the superior portion of the crest, which, a little below that tuberosity, turns abruptly backwards, and thus describes a curve the concavity of which is directed forward. The superior part of the external surface is very much hollowed out.

The superior extremity is much thicker than the inferior one. It is not only wide in the transverse direction, but is more especially extended from before backwards; the prominence of the anterior tuberosity is the cause of the elongation of this antero-posterior diameter. On the posterior part of the external tuberosity is found a surface to which the superior extremity of the fibula is applied.

The inferior extremity presents an articular surface, which is formed of two lateral cavities, separated by a crest, which is directed obliquely forwards and outwards. The internal part is prominent, and forms the internal malleolus.

With regard to the fibula, it is united to the tibia by its extremities and by the inferior half of its shaft. This latter is more expanded below than in its upper part. The superior extremity is flattened from without inwards. The inferior extremity projects beyond the articular surface of the tibia, and forms the external malleolus, which, instead of, as in man, descending further than the tibial malleolus, stops at the same level, and even descends a little less than does the latter.

In the cat, the curve of the tibia is less pronounced; this is due to the fact that the crest, instead of being concave in its middle portion, is slightly convex anteriorly. The fibula, less flattened than that of the dog, is united to the tibia by its extremities only, and is separate in the rest of its extent.

The bones of the tarsus are seven in number, and arranged as in man, with this difference (which is easily comprehended), that their general relations are changed on account of the vertical direction of the tarsus. For example, the astragalus, instead of being above the calcaneum, is situated in front of it; the cuneiform bones, instead of being situated in front of the scaphoid, are found below it, etc.

These animals have but four well-developed metatarsals; that which corresponds to the great-toe is represented merely by a small style-shaped bone, situate at the internal part of the region.

Nevertheless, we find this toe fully developed in some dogs. Notwithstanding this, the bones which form it are, however, but rudimentary, and much smaller than those of the innermost digit of the fore-limb.

Sometimes it is double; this condition is demonstrable in individuals belonging to breeds of large size. The median metatarsals are more fully developed than the other bones of the same region which are next them. Viewed as a whole, the metatarsal bones are a little longer than the metacarpals; the result is that the distance which separates the tarsus from the ground is a little greater than that which separates the carpus from the plane on which the anterior limbs rest. The length of the calcaneum still further exaggerates this difference, and, as in the animals with which we shall occupy ourselves later on, the projection which this bone forms is distinctly higher than that which is produced by the pisiform.

The metatarsus, as a whole, is a little narrower than the metacarpus; not only on account of the presence of a thumb in the anterior limb, but, further, because the bones of this latter region are wider than those of the corresponding part of the posterior limb.

The phalanges closely resemble those of the anterior limbs.

Unguligrades: Pig (Fig. 38, p. 58).—The pelvis in this animal presents a few of the characters which we shall again meet with in the ruminants and the solipeds; however, the posterior (or internal) iliac spines are relatively more widely separated from one another than in the latter. This arrangement reminds us of that found in the carnivora.

The femur presents nothing very special. The knee-cap is thick, and ovoid in outline.

The fibula is completely developed, as in the carnivora; and is connected with the tibia at both its extremities.

The tarsus consists of seven bones. The astragalus and the calcaneum differ slightly from those of ruminants.

The foot, like the hand, has two median digits which rest on the ground by their third phalanges; and an internal and an external digit, which are removed from it. The metatarsals are a little longer than the metacarpals.

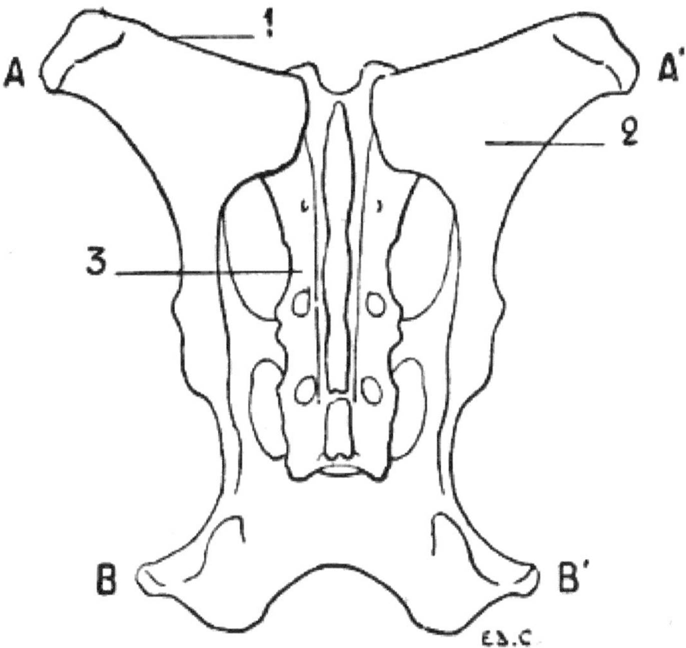

Fig. 51.—Pelvis of the Ox: Superior Surface.

1, Iliac crest; 2, external iliac fossa; 3, sacrum; AA′, bi-iliac diameter; BB′, bi-ischiadic diameter.

Unguligrades: Sheep, Ox (Fig. 39).—The pelvis of ruminants of this group closely resembles that of the horse, which we will study later on. That which we must at once point out is that, with regard to the ratio formed by a comparison of the bi-iliac and bi-ischiatic diameters, it may be placed between the ratio obtained in comparing those diameters in the pelvis of the carnivora and that of the solipeds. Indeed, in the ruminants, the distance which separates the ischia exceeds the width of one iliac only, and does not equal, as in the felide, the total width of the anterior part of the pelvis (Fig. 51). In the skeleton of the ox, which forms part of the anatomical museum of the École des Beaux-Arts, the bi-ischiadic diameter is 39 centimetres, whilst the width of one iliac crest is 29 centimetres, so that, in contrast to that which we find in the dog, the width of the ischiadic region is less than that formed in front by the addition of the iliac crests.

The great trochanter is large, and extends beyond the level of the plane in which the head of the femur is found.

In the ox, the linea aspera, instead of being a narrow crest, is spread out, and forms in reality a surface; the posterior surface of the femur. At the inferior and external part of this surface is situated a cavity which surmounts the corresponding condyle, and is known as the *supracondyloid fossa*. On the internal part of the same region there are a series of tubercles, which, because of their position in relation to the corresponding condyle, constitute the *supracondyloid crest*.

The internal lip of the trochlea is much thicker and much more prominent than the external.

The details which we have just now examined in connection with the ox are less marked in the sheep.

The trochlea, narrow as a whole, is clearly separate from the condyles by a very marked constriction.

The patella, which is thickened in the antero-posterior direction, has the shape of a triangular pyramid with the base upwards. Its posterior surface, which articulates with the trochlea, presents an arrangement which is adapted to the disposition of this latter—that is to say, the surface which is in contact with the internal lip is larger than that which articulates with the lip of the opposite side.

The tibia of the ox is proportionately shorter than that of the sheep. The shaft of this bone is flattened from before backwards, in its inferior half. The median crest of the articular surface of the inferior extremity is the most prominent part of that region.

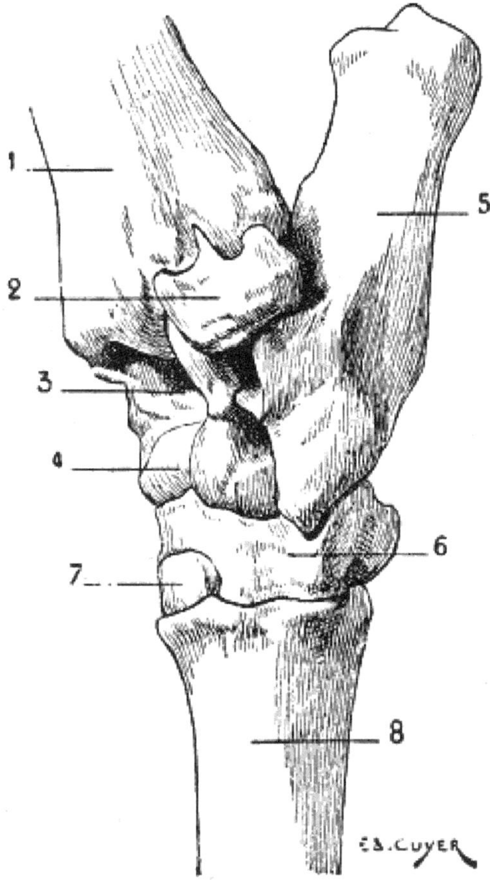

Fig. 52.—Tarsus of the Ox: Posterior Left Limb, Antero-external Surface.

1, Tibia; 2, coronoid bone of the tarsus; 3, superior articular surface of the astragalus; 4, inferior articular surface of the astragalus; 5, calcaneum; 6, cuboido-scaphoid bone; 7, great cuneiform bone—the small cuneiform bone is situated at the back of the latter; 8, principal metatarsal—the small, or rudimentary, metatarsal bone is very small; it is situated at the back of the preceding, and is not to be seen in the sketch. It would be visible if the view were directly lateral, but then the superior and inferior articular surfaces of the astragalus would be less apparent.

The fibula is extremely atrophied. The shaft and superior extremity of this bone are represented merely by a simple ligamentous cord, which is sometimes ossified. There remains of the fibula, as a portion well and distinctly developed, the inferior extremity only. This presents itself under the form of a small bone situated in the region ordinarily occupied by the inferior extremity of the outer bone of the leg—that is to say, the external part of the inferior extremity of the tibia; this little bone articulates with the astragalus and the calcaneum. Some authors consider it to be a tarsal bone, and describe it under the name of the coronoid bone of the tarsus (Fig. 52, 2). It is not, perhaps, quite legitimate to describe it as a bone of this region, for it has not a homologue in the tarsus of other animals. Its external surface is rough; its superior border is furnished with a small pointed process occupying a depression which is provided for it by the tibia. It reaches lower down than the latter, and forms in this way a sort of external malleolus, which frames, on the outer aspect, the mortise in which the astragalus is maintained.

The tarsus, as a whole, has an elongated form; it is formed of five bones: the astragalus, calcaneum, cuboid and scaphoid, which coalesce, to form a single bone, and two cuneiform bones, which correspond to the second and third cuneiform bones of the human foot. These cuneiforms are called, from their size, commencing internally, by the names small and great cuneiform.

The calcaneum is long and narrow; it is longer than that of the horse; it is on the anterior and external part that the bone (coronoid tarsal bone) which represents the inferior extremity of the fibula is situated. It forms the prominence known as *the point of the ham*, a prominence which is no other than the heel, which, in the unguligrades, is, as we have already said, very far removed from the ground.

The astragalus, which is elongated in the vertical direction, has three articular surfaces disposed in the form of trochleæ: a superior trochlea, which is in contact with the skeleton of the leg, and which is present in all animals; an inferior, which replaces the articular head found on the anterior aspect of the astragalus in man; this articulates with the portion of the scaphoido-cuboid that corresponds to the scaphoid; and, lastly, a posterior trochlea with which the calcaneum articulates. Of these three trochleæ, the superior is the most strongly marked. Between this latter and the inferior is found, on the anterior surface of the astragalus, a deep depression, which, during flexion of the foot on the leg, receives a prominence which the inferior extremity of the tibia presents in its median portion.

We can easily recognise the trochleæ which we have been discussing, in the little bones which children use 'to play at bones'; these bones are no other than the astragali of sheep.

We have already mentioned that the scaphoid and the cuboid are ankylosed; they form by their union an irregular bone, on which the astragalus and calcaneum are supported.

The cuneiforms articulate with the internal half of the superior extremity of the principal metatarsal; the external half of this metatarsal articulates with the portion of bone which represents the cuboid.

The metatarsus is represented by a principal metatarsal, formed by the coalescence of two metatarsals; we also find in this region a very small rudimentary metatarsal.

The metatarsus is a little longer than the metacarpus; its transverse measurement is a little less; on the other hand, it is a little thicker in antero-posterior direction; from these two differences it results that the body of the metatarsus is quadrilateral, whereas the metacarpus presents only an anterior and a posterior surface.

The rudimentary metatarsal is a very small roundish bone, situated at the back of the superior extremity of the principal metatarsal.

The phalanges closely resemble those of the anterior limbs; nevertheless, the first and second phalanges differ from the latter in the fact that they are a little longer and narrower.

At the back of the metatarso-phalangeal articulations, as in the corresponding region of the anterior limbs, are found the sesamoid bones. Such also exist at the articulations of the second and third phalanges.

Unguligrades: Horse (Fig. 40).—The pelvis of the horse presents a general form which sharply differentiates it from that of the carnivora; in fact, the ilium is twisted in such a way that the external iliac fossa does not look outwards, but upwards. It results from this twist that the anterior iliac spine, which we have seen to be directed downwards in the carnivora, has become external; and this prominence is much farther removed from the vertebral column than in the dog or cat. On the other hand, the posterior iliac spine, which is directed upwards in the carnivora, has become internal; it is also placed nearer to the vertebral column, with the result that the distance which separates this spine from that of the bone of the opposite side is proportionately less.

The internal iliac spine, which is conical in shape, and curved upwards, forms a prominence known as *the angle of the crupper*; the external iliac spine, thick and provided with tuberosities, forms a clearly-defined prominence; this is the angle of the *haunch*.

The iliac crest, extending directly from one spine to the other, is curved, its concavity being turned upwards. The external iliac fossa, which looks upward, is limited anteriorly by this crest, and is, like the latter, slightly hollowed. The portion of the bone which connects the ilium to the region occupied by the cotyloid cavity is extremely narrow; posteriorly, the bone enlarges again to form the ischial and pubic portions.

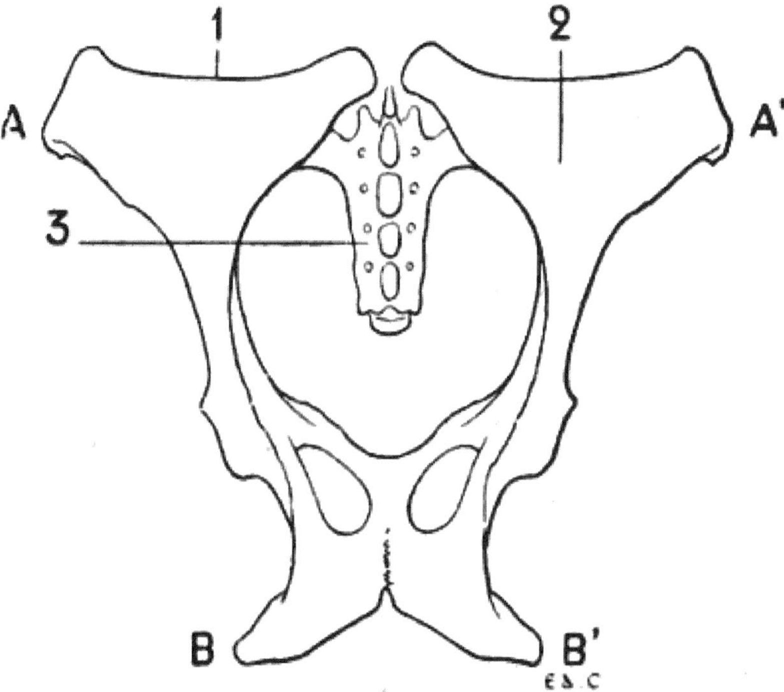

Fig 53.—Pelvis of the Horse: Superior Surface.

1, Iliac crest; 2, external iliac fossa; 3, sacrum; AA′, bi-iliac diameter; BB′, bi-ischiatic diameter.

The tuberosity of the ischium, thick and curved upwards, but less so than in the ox, forms the most prominent part of the posterior border of the region of the thigh; this projecting portion, so sharply defined in spare subjects, is known as the *point* or *angle of the buttock*. Contrary to what we have indicated in the case of the dog, the distance which separates the ischiatic tuberosities is inconsiderable in proportion to that which we find between the external iliac spine of one side and that of the opposite. The bi-ischiatic diameter does not even equal the width of one iliac bone measured at the level of its crest (Fig. 53). On the skeleton of the horse in the École des Beaux-Arts, the distance which separates the tuberosities of the ischia is 225 millimetres; that between the two spines of each iliac bone is 25 centimetres.

The anterior region of the crupper is thus much broader than that occupied by the ischia.

The femur is relatively short. Its shaft is rectilinear, and does not present the anterior convexity which is found on the human femur, and which we indicated when discussing that of the dog. The shaft of the bone, instead of being prismatic and triangular, presents four surfaces; the anterior, internal, and external, almost pass into each other, being separated one from the other merely by rounded and slightly marked borders; the posterior surface, which is plane, replaces the linea aspera, which in the horse, instead of presenting the appearance of a crest, is considerably widened. The numerous irregularities which this surface presents give insertion to the muscles which correspond to those attached to the linea aspera.

Between this posterior surface and the external is found a rough prominence which curves forward; this was designated by Cuvier the *third trochanter*; it replaces the external branch of the superior line of bifurcation of the linea aspera; other authors call it the *infratrochanteric crest*, because it is situated below the great trochanter. At the inferior part of the same region is found a deep fossa, the borders of which are rough; this is the *supracondyloid fossa*.

Between the posterior surface and the internal are found: above, the lesser trochanter, which is long and rough; below, at the level of the supracondyloid fossa, an equally rough surface known by the name of the *supracondyloid crest*.

The superior extremity is flattened from before backwards. The neck is not well marked. The great trochanter is very prominent, and projects beyond the level of the head of the femur. We divide the great trochanter into three parts: the summit, which is the most elevated portion; the convexity, which is situated in front; and the crest, formed by muscular impressions, situated outside and below the convexity. The digital fossa is situated behind and below the summit of the great trochanter. With regard to the lesser trochanter, it is placed so far down that it really forms part of the shaft of the bone, with which, besides, we have described it.

On the inferior extremity of the femur are two condyles and a trochlea; the condyles are clearly separated from this latter by a marked constriction.

The trochlea is directed with a slight obliquity downwards and inwards; its internal lip is much thicker and more prominent than the external; this is, accordingly, a condition exactly the opposite of that which characterizes the corresponding region of the human femur.

The knee-cap is lozenge-shaped; its superior angle projects upward, and produces a prominence at the part which corresponds to the base of the human patella, the part which is here the thickest portion of the bone. Its anterior surface is convex and rough. Its posterior surface presents two lateral articular facets, separated by a crest; this surface is in contact with the trochlea of the femur, and, as it is the internal lip of the latter which is the more developed, it results therefrom that the internal articular surface of the knee-cap is larger than the external.

The knee-cap contributes to the formation of the region of the posterior limb which is called the *stifle*.

The tibia is large in its upper portion; in its inferior part it is flattened from before backwards. The posterior surface of the shaft presents an oblique line, below which are found vertical rough lines for the insertion of muscles. The external surface is hollowed out in its upper part. The anterior tuberosity of the tibia rises just to the level of the flat articular surface; it is hollowed in its median portion by a vertical groove of elongated form, which receives the ligament that binds the knee-cap to the tibia. The external tuberosity is more prominent than the internal; in it is found a groove for the passage of the anterior tibial muscle.

The inferior extremity, flattened from before backwards, presents a surface which is moulded on the trochlea of the astragalus; the median crest of this surface is thick, and descends lower posteriorly than the tuberosities which are situated on the external and internal aspects of this extremity.

Of the two tuberosities, that which is internal is comparable to the internal malleolus of man, the one on the outer side forms a sort of external malleolus; but this latter here belongs to the tibia, and not to the fibula.

The fibula, in fact, does not reach the inferior extremity of the tibia; it is a poorly developed bone, elongated and terminating inferiorly in a point, at the middle of the shaft of the tibia or at its lower third. Its superior extremity, which is slightly expanded, articulates with the tuberosity which occupies the outer aspect of the corresponding extremity of the tibia.

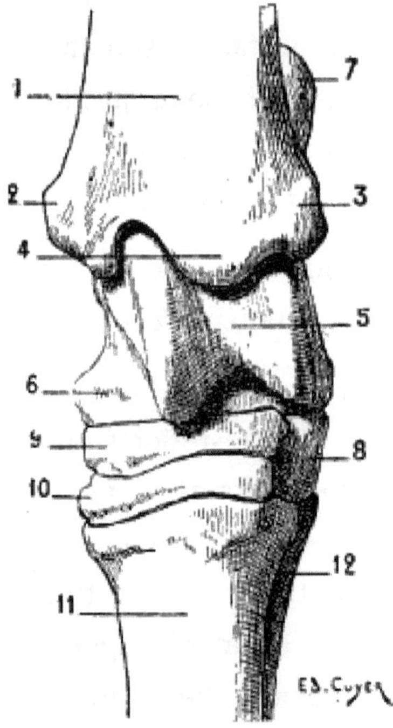

Fig. 54.—Tarsus of the Horse: Left Posterior Limb, Anterior Surface.

1, Tibia; 2, internal tuberosity of the inferior extremity of the tibia (homologue of the internal malleolus of man); 3, external tuberosity of the inferior extremity of the tibia (homologue of the external malleolus); 4, median crest lodged in the groove of the pulley of the astragalus; 5, pulley of the astragalus; 6, internal tuberosity of the astragalus; 7, calcaneum; 8, cuboid; 9, scaphoid; 10, great cuneiform, the small cuneiform is placed behind this latter; 11, principal metatarsal; 12, external rudimentary metatarsal. The internal rudimentary metatarsal, being more slender than the external, does not appear in the figure.

The bones of the tarsus are six in number: the calcaneum and astragalus form the upper row; the cuboid, scaphoid, and two cuneiforms form the lower (Fig. 54).

The astragalus has not, as in ruminants, an inferior trochlea for articulation with the scaphoid; this portion of the bone presents a surface which is slightly convex. It articulates with the tibia by a trochlea that occupies not only the superior surface, but also the anterior. This trochlea, which is directed slightly obliquely downwards and outwards, has a very pronounced form; its lips, which are extremely prominent, determine by their anterior part one of the features which we recognise on the anterior aspect of the *ham*—a feature which is still more accentuated when the metatarsus (*canon*) is extended on the leg. On the internal surface of the astragalus is found a tubercle, which forms a projection in the corresponding region of the ham.

The calcaneum, which is not quite so long as that of the ox, forms by its summit a prominence which is called *the point of the ham*.

The cuboid is small; the scaphoid is large, and flattened from above downwards. Of the two cuneiforms, the more external is the larger; it closely resembles the scaphoid; it is flattened from above downwards as is the latter; but it is a little smaller in size. The small cuneiform, which occupies the inner side of the tarsus, is the smallest bone in this region; it is sometimes divided into two parts; this raises the number of the cuneiforms to three, and that of the bones of the tarsus to seven.

The bones of the metatarsus and the phalanges are equal in number to the corresponding bones in the anterior limbs; they are formed on a type analogous to that of these latter. Accordingly, we shall merely indicate the differences which characterize them.

The principal metatarsal is longer than the metacarpal of the same class; its shaft is more cylindrical; its inferior extremity is somewhat thicker. The external rudimentary metatarsal is better developed than the internal; in the metacarpus the reverse is the case.

The phalanges so far resemble those of the anterior limb that, as differential characters, we need point out only the following: the first phalanx of the hind-foot is a little shorter than that of the fore-foot; its inferior extremity is a little narrower, and its superior extremity a little thicker. The second phalanx is a little less expanded laterally.

The difference in appearance which the three phalanges, anterior and posterior, respectively present are to be borne in mind; for they are correlated to the general form of the fore and hind feet. We will establish this point when we come to study the hoof (see Figs. 101 and 102). In the fore-foot the ungual phalanx has its inferior surface limited externally by a circular border, while the same bone of the hind-foot has this surface a little narrower, more concave, and limited by two curved borders which unite anteriorly to form an angle—an arrangement which gives to the general outline of this region the form of the letter V.

Articulation of the Posterior Limbs

The Coxo-femoral Articulation.—The head of the femur is received in the cotyloid cavity; these are the osseous surfaces in contact in this articulation. They are maintained in position by a fibrous capsule and a round ligament. To this latter is found attached, in the horse, a fasciculus which, commencing, as does the round ligament, at the depression on the head of the femur, emerges from the cotyloid cavity by the notch which is present in its circumference, and is attached to the anterior border of the pubes, to blend with the tendon of the rectus muscle of the abdomen. This is the pubio-femoral ligament.

The movements which this joint permits are the same in the quadrupeds as in man, but less extensive. They are: flexion and extension, abduction and adduction, the two latter being much more limited than the former. There is also rotation.

By flexion, the inferior extremity of the femur is directed forwards; the bone of the thigh then takes a more oblique direction than the normal. This movement takes place, for example, when the animal carries forward one of its hinder limbs. Extension, which takes place in an inverse sense, is produced when the foot is fixed on the ground, while the body is projected forward. It is also produced in the action of kicking.

As for the lateral movements—viz., abduction and adduction—they are less extensive than the preceding movements. The absence of the pubio-femoral ligament in other quadrupeds than the horse explains why in them abduction is less limited than in the latter. Indeed, it is the tension of this ligament, occasioned by the abduction of the thighs, which arrests more quickly the movement in question.

Articulation of the Knee.—This articulation, as in man, is formed by the femur, the patella, and the tibia.

In the horse the ligament of the patella is not single, but consists of three parts, designated, on account of their position, by the respective names of external, internal, and median patellar ligaments. The two former come from the angles on the corresponding borders of the knee-cap; the median springs from the anterior surface and inferior angle of the same bone. They all three pass to their termination on the anterior tubercle of the tibia. The external ligament is the strongest, and the internal ligament the least developed.

In the dog, the cat, the pig, and the sheep, the patellar ligament consists of a single band. The articulation is further strengthened on the sides by lateral ligaments—an internal and an external.

With regard to the principal movements, these are flexion and extension, to which may be added movements of rotation of limited extent. In flexion, the leg bends on the thigh; its inferior extremity is directed upwards and backwards; the angle which the tibia naturally forms with the femur becomes less obtuse.

But it should be understood that one part of this description—that which has relation to the leg—holds good only when the femur is in its normal condition, or in flexion. Indeed, at the close of the movement in which, during a step, the foot is in contact with the ground—that is, at the termination of the resting stage—the inferior extremity of the tibia is directed backwards. But the femur is then in a state of extension, and in regard to this latter the attitude of the leg is unchanged.

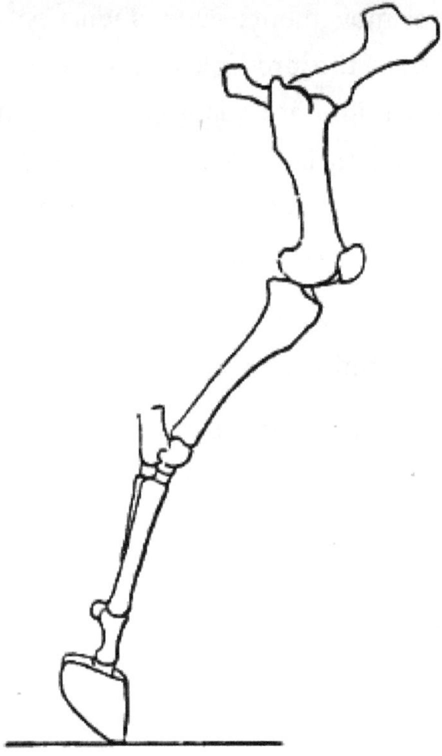

Fig. 55.—Extension of the Leg: Right Posterior Limb of the Horse, External Surface. (After a Chronographic Study by Professor Marey.)

At this moment, notwithstanding the direction, which recalls that which it has at the time of flexion, the leg is not bent on the thigh; on the contrary, it is almost in the line of its continuation (Fig. 55). As we have done in connection with the articulations of the anterior limbs, we borrow this figure from the interesting chronophotographic studies of Professor Marey.[17]

[17] E. J. Marey, 'Analysis of the Movements of the Horse by the Chronophotograph' (*La Nature*, June 11, 1898).

The Tibio-tarsal Articulations and of the Bones of the Tarsus.—In the region which veterinary anatomists call the ham, the articulations of the leg and foot alone call for special study in the case of the horse. The articulations of the bones of the tarsus, and of these with the metatarsus, do not offer any interest with regard to mobility, this being almost wholly absent at that level.

The leg and the astragalus, in a general way, are placed in contact by such articular surfaces that the resulting joint, which is a true hinge, permits movements of flexion and extension only. Indeed, as we have indicated above, the tibia is furnished, on the inferior surface, with a crest that fits into the deep groove which is situated on the corresponding surface of the astragalus.

During flexion, the anterior surface of the foot tends to approach the anterior surface of the leg, the angle formed by these two segments becoming more and more narrowed. The displacement in the opposite direction characterizes extension.

In other quadrupeds, the articulations which bind together the bones of the tarsus possess a little more freedom of movement. The shape of these bones, and particularly the shape of the surfaces of the astragalus, which are in contact with them, allow movements in this region, in the case of the dog and cat, which, without being so extensive as those of the human foot, in the subastragaloid articulation, nevertheless, recall the mobility which we find in the human species at this level—that is to say, rotation, abduction, and adduction of the foot.

As for the articulations of the metatarsus with the phalanges, and of the phalanges with one another, they resemble those of the anterior limb too closely that it should be necessary to study them here. Such a study would be, in this case, but a repetition (see, a description of the articulations in question).

THE HEAD IN GENERAL, AND IN SOME ANIMALS IN PARTICULAR.

When we compare, by the examination of one of their lateral aspects, the skull of man and the same region in other mammals, it is easy to observe that the relative development of the cranium and face is entirely different. In the case of man the cranium is large, and the face relatively small; in animals the face is proportionally much more highly developed. The measure of the facial angle permits us to note these differences, and the figures relative to the value of this angle are sufficiently demonstrative to induce us to indicate those which are, in a general way, connected with some of the forms in individuals which here occupy our attention. In the first place, we must remember that the angle in question is more acute, as the cranium is less developed in proportion to the facial region (Figs. 56 and 57). It is especially to this character that we wish to draw attention.

Man	70° -80°
Cat	41°
Dog	28° -41°
Sheep	20° -25°
Ox	18° -20°
Ass	12° -16°
Horse	11° -13°

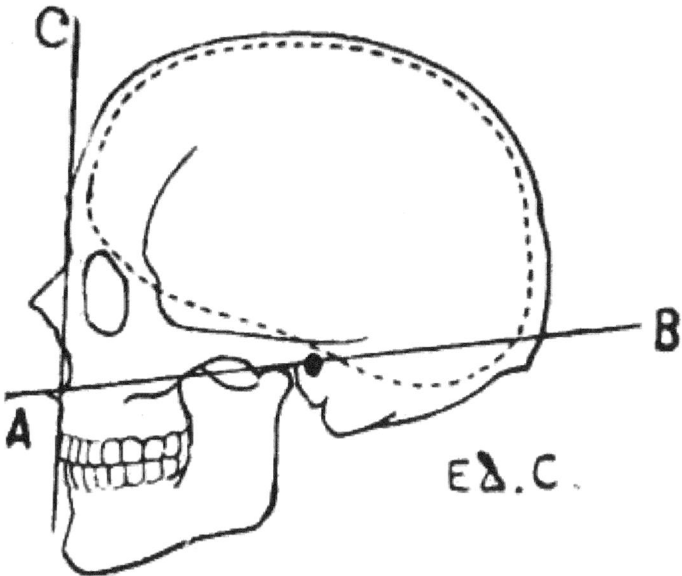

Fig. 56.—Human Skull: Measure of the Facial Angle by the Method of Camper. Angle BAC = 80°.

The internal wall of the cranial cavity is marked by the dotted line.

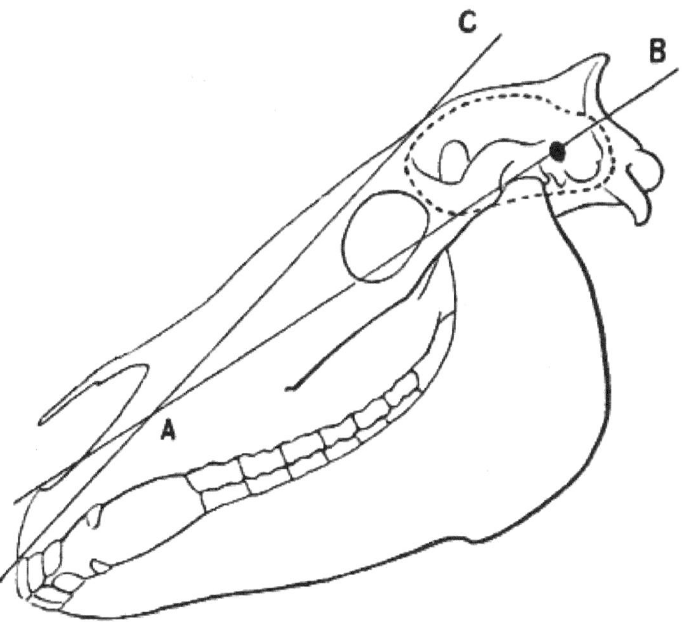

Fig. 57.—Skull of the Horse: Measure of the Facial Angle by the Method of Camper.[18] Angle BAC = 13°.

The internal wall of the cranial cavity is shown by the dotted line.

[18] We have indicated on this sketch of the skull of the horse the facial angle measured by the method of Camper, in order that the correspondence with Fig. 56 may be more complete. But it is certain that the procedure here employed is in practice not satisfactory, since the apex of the angle, as we can demonstrate, is found to be situated within the contour of the head, and that, consequently, it is rather difficult to localize it precisely in the case of a given skeleton. Further, because of the absence of the base of the nose in the complete skull, the auriculo-nasal line cannot be accurately fixed. It would be the same for most other animals. This is why the method employed for these latter is preferably that of Cuvier, or, again, that of Cloquet. In the former, the apex of the angle of Camper is transferred to the free border of the upper incisors, but these teeth may be absent, and, on the other hand, ruminants are destitute of them. In the second, the same apex is placed at the alveolar border, and the angle then becomes fairly easy to appreciate.

Besides, in animals the cranium is very prominent superiorly, and the face, more or less elongated, is sharply projected downwards and forwards; in man the cranial region occupies not only the superior, but also the posterior part; the face is short and of a compact form. The human head, in its general aspect, may be compared to a sphere, while the skull of the quadrupeds presents the aspect of a quadrangular pyramid, with the base turned upwards and the summit at the incisor teeth.

Direction of the Head.—Before entering on the study of the bones of the head, it is necessary, in our opinion, to agree as to the position in which we shall suppose it to be placed.

The question may seem to be one of little importance; nevertheless, it cannot be regarded as indifferent, since authors are not all agreed on this subject.

Some suppose it to be placed vertically—that is, with the incisor teeth turned directly downwards. Others, on the contrary, suppose it to be placed horizontally, resting on the whole length of the lower jaw, the face being then turned upwards. These two extreme methods of arrangement appear to us to possess inconveniences—at least, for comparison with the human head.

Indeed, if, when the head is vertical, the same regions of the face (forehead, nose) are, in the case of animals as well as man, turned forward, the lower jaw ceases to merit its appellation, as it is then situated, not below, but behind the upper. Furthermore, if this position is chosen, for example, for modelling or drawing, it cannot be obtained without difficulty when we have to deal with an isolated piece of the skeleton, on account of the absence of equilibrium, which it is necessary to obviate. It is true that the question of convenience should not take precedence of all others, and it suffices for us in this connection to recall, in regard to the human pelvis, that, although the older anatomists used to represent it as resting commodiously on the three angles which terminate it at its lower part (ischial tuberosities and coccyx), this attitude being false, it is customary now to incline the superior aspect forwards, inasmuch as this arrangement more nearly conforms to reality, in spite of the fact that it is a little more difficult so to dispose an isolated pelvis. Further, to return to the head; if its vertical direction can be demonstrated, for example, in many horses, it is not sufficiently general to be adopted as the classic position.

In regard to the facility of placing in position, the horizontal direction is certainly to be preferred; but this is also far removed from the natural position in the animal while in the state of repose. On the other hand, the mind is not satisfied with the idea that certain regions of the face, such as the nose and the forehead, are then directed upwards. And yet it is necessary to come to a decision, seeing that what we are now investigating applies also to the position to which it is necessary to give the preference in placing the skeleton of the head when we wish to draw it in profile. That which we adopt is a compromise, but to us it seems more rational.

The position of the head of the horse, to be normal, should be such as to give it an inclination of 45°. In this case the lower jaw is still posterior; and, for this reason, we see in adopting this position some inconveniences from a didactic point of view. Accordingly, we will suppose the head brought a little nearer to the horizontal, and this, from the imaginative point of view, has certainly an advantage which we cannot afford to neglect when addressing artists.

Indeed, let us suppose that to a clay model of a human head we wish to give the aspect of the head of a quadruped. We should elevate the occiput; and then, taking hold of the lower part of the face, we should lengthen it, not in a direction precisely antero-posterior, but downwards and forwards. It is obviously this latter procedure which, on the other hand, is carried out when a person wishes to give to his own face some resemblance to the muzzle of a quadruped.

It is true that, in the position we have adopted, the face is directed obliquely downwards and forwards, and that there may result a certain confusion in describing the position of its different parts. On this account, with the object of not making complications, we purpose, for the present, to substitute, for example, for the term 'antero-superior'—which when speaking of the position of the forehead and nose would be more exact—the term 'anterior,' which is sufficiently comprehensible. The mouth will be, for the same reason, referred to as being situated at the inferior part of the face, and not the antero-inferior.

The Skull.—The elevation of the cranial region becomes especially appreciable when we examine the occipital bone. Before verifying this fact, it is not superfluous to recall the general arrangement which this bone presents in the human skull. A portion of the occipital bone occupies the base of the skull; but this base in man is horizontal; to this region succeeds the shell-shaped portion of the occipital bone, which, passing vertically upwards, forms with the preceding portion an angle situated at the level of the external occipital protuberance, and of the curved line which starts from it on each side. In animals a portion of the occipital bone is horizontal, it is true; but this bone being sharply bent at the level of the occipital foramen and condyles, the result is that the portion which surmounts these latter looks backwards, and is limited above by the external occipital protuberance, which forms the culminating point of the skull; this point is situated between the ears.

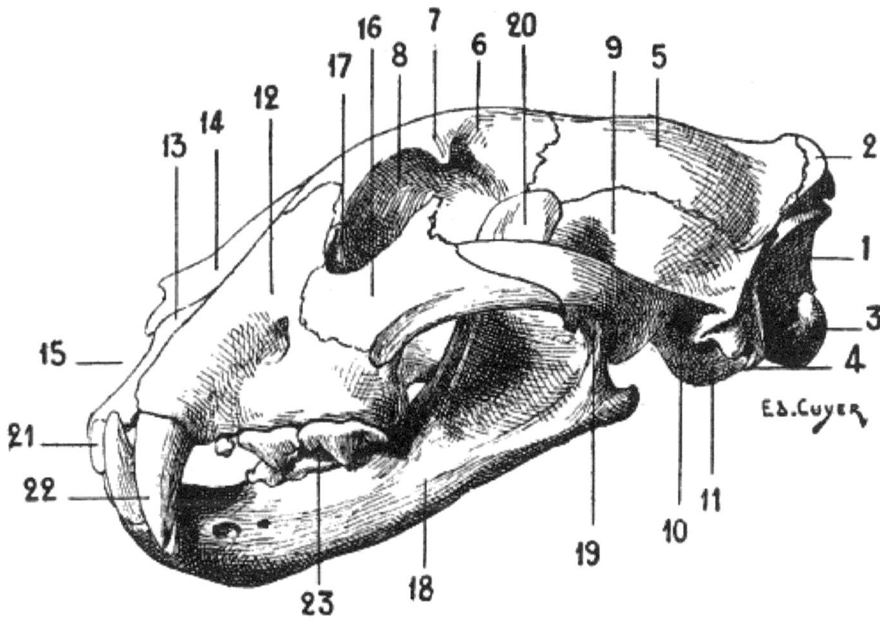

Fig. 58.—Skull of one of the Felidæ (Jaguar): Left Lateral Aspect.

1, Posterior surface of the occipital bone; 2, external occipital protuberance; 3, condyle of the occipital bone; 4, jugular process; 5, parietal bone; 6, frontal bone; 7, orbital process; 8, orbital cavity; 9, squamous portion of the temporal bone; 10, external auditory canal, in front of which is situated the zygomatic process; 11, tympanic bulla; 12, superior maxillary bone; 13, intermaxillary or incisor bone; 14, nasal bone; 15, anterior orifice of the nasal cavity; 16, malar bone; 17, ungual or lachrymal bone; 18, inferior maxillary bone; 19, condyle of the inferior maxillary bone; 20, coronoid process; 21, incisor teeth; 22, canine teeth; 23, molar teeth.

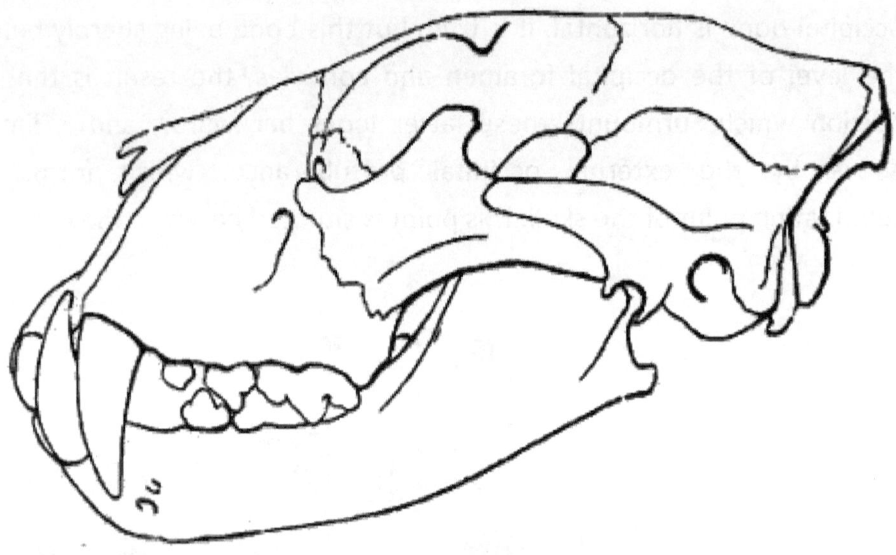

Fig. 59.—Skull of the Lion: Left Lateral Aspect.

This figure is intended to show that in the lion the contour of the face between the nasal bones and the cranial region is more flattened than in other felidæ, such as the tiger, jaguar, panther, and domestic cat. This difference is shown by comparison of this figure with the preceding one (Fig. 58). We are indebted to M. Tramond, the well-known naturalist, for the indication of this differential character which, from the artistic plastic point of view, is one of real interest.

This protuberance, prolonged on each side by the superior curved line of the occipital bone, is so much the more prominent as this bone bends sharply a second time, so as to form a third portion, which, looking forwards, forms part of the anterior aspect of the skull, and proceeds to articulate with the parietals. On this third portion is found a crest which, proceeding from the occipital protuberance, is continuous in front with the parietal crests, to which we will again refer in speaking of the parietal bones.

On the inferior surface of the human occipital bone are found, at the level of, and external to, the condyles two bony elevations which bear the name of *jugular eminences*. They are long in quadrupeds, and constitute what are designated by some authors the *styloid processes*, but they must not be confounded with the processes of the same name which in the case of man form part of the temporal bone. These processes are very highly developed in the pig, horse, ox, and sheep.

In the ox, the occipital bone is deprived of the protuberance, and is not bent on itself in the anterior portion, neither does it form the most salient part of the skull; this latter, which is situated at the level of the horns, belongs to the frontal bone. In the pig, also, the occipital bone is not bent upon itself in its anterior portion, but forms the summit of the head. The occipital protuberance, hollowed on its posterior surface, rises vertically, and rests upon the parietal bone, with which it forms an acute angle.

The parietals, two separate bones in the dog and the cat, but fused in the median line in the ox, sheep, and horse, are of special interest in regard to the two crests which, in the carnivora, and also in the pig and the horse, occupy their external surface, and, after diverging from one another, are continued by a crest which crosses the frontal bone and ends at the external orbital process of the latter bone.

These crests, known as the *parietal* or *temporal crests*, recall both in position and relations the temporal curved line of the parietal bone of man. They contribute, as in the case of the latter, to the formation of the boundaries of the temporal fossa.

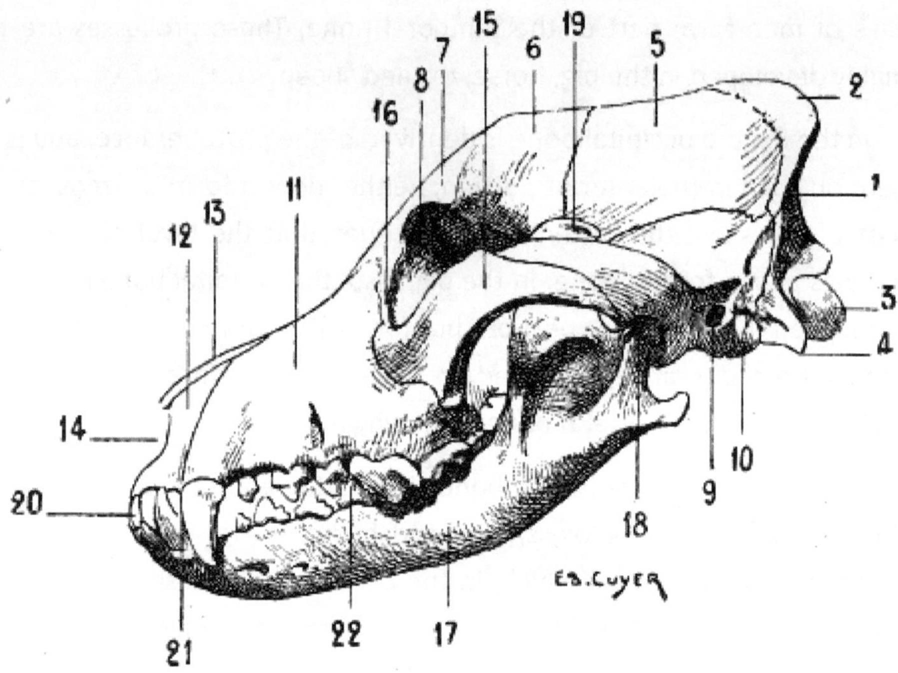

Fig. 60.—Skull of the Dog: Left Lateral Aspect.

1, Posterior surface of the occipital bone; 2, external occipital protuberance; 3, occipital condyle; 4, jugular process; 5, parietal bone; 6, frontal bone; 7, orbital process; 8, orbital cavity; 9, external auditory canal, in front of which is found the zygomatic process; 10, tympanic bulla; 11, superior maxillary bone; 12, intermaxillary or incisor bone; 13, nasal bone; 14, anterior opening of the cavity of the nasal fossæ; 15, malar bone; 16, lachrymal bone; 17, inferior maxillary bone; 18, condyle of the inferior maxillary bone; 19, coronoid process; 20, incisor teeth; 21, canine teeth; 22, molar teeth.

In the carnivora, these crests are situated, throughout their whole length, in the median line, the temporal fossæ being, accordingly, as extended as they possibly can be. In certain species, the development of these crests is such that they form by their union a vertical plate, which, in separating the two temporal fossæ, gives them a greater depth. In the pig, the parietal crests, analogous in this respect to the temporal curved lines of the parietal bones of man, are separated by an interval, proportionately less extended, however, than that of the human skull. The parietal bone in the ox and the sheep does not enter into the formation of the anterior surface of the skull; it is formed by an osseous plate, narrow and elongated transversely, which, with the occipital bone, constitutes the base of the region of *the nape of the neck*. It is bent upon itself at the level of its lateral portions so as to occupy the temporal fossa.

The anterior surface of the frontal bone, which is depressed in the median line in the dog, but plane in the horse, is limited by two crests, which, situated on the prolongation of the parietal crests, diverge more and more from one another in proportion as they occupy a lower position. This surface terminates externally in two processes, which are the homologues of the external orbital processes of the human frontal bone.

The superior border of these orbital processes, situated on the prolongation of the corresponding parietal crests, contributes to limit the temporal fossa. Each of these orbital processes terminates in the following manner: In the bear, dog, cat, and pig, in which the orbital cavities are incompletely bounded by bone, this process, slightly developed, is not in connection, by its inferior extremity, with any other part of the skeleton of the region. In the ox and the sheep, it articulates with a process of the malar bone. In the horse, it articulates with the zygomatic process of the temporal bone. The inferior margin of this process forms a part of the boundary of the anterior opening of the orbital cavity.

The supra-orbital foramen, which does not exist in carnivora, occupies in the horse the base of the orbital process. In the ox, it is situated a little nearer the middle line; and its anterior orifice opens into an osseous gutter which is directed upwards towards the base of the horn, while inferiorly it meets the inferior border of the frontal bone; in the sheep this groove is but slightly developed. In this latter, as in the ox, it is the frontal bone which forms the most elevated portion of the skull. In fact, being bent upon itself at a certain level, its external surface is formed of two planes: one, posterior, which is inclined downwards and directed backwards; the other, anterior, is also inclined downwards, but with a forward obliquity. At the union of these planes the bone forms an elbow, on either side of which are found the osseous processes on which the horns are mounted.

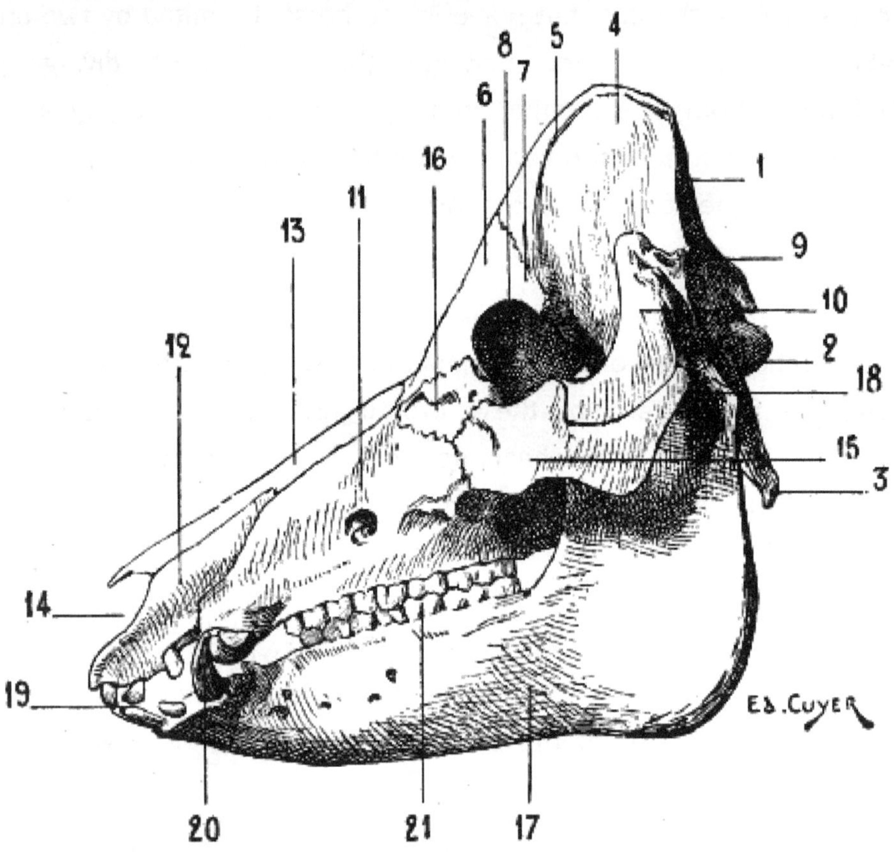

Fig. 61.—Skull of the Pig: Left Lateral Aspect.

1, Occipital bone; 2, condyle of the occipital; 3, jugular process; 4, parietal bone; 5, parietal crests; 6, frontal bone; 7, orbital process; 8, orbital cavity; 9, external auditory canal; 10, zygomatic process; 11, superior maxillary bone; 12, intermaxillary or incisor bone; 13, nasal bone; 14, anterior orifice of the cavity of the nasal fossæ; 15, malar bone; 16, lachrymal bone; 17, inferior maxillary bone; 18, condyle of the inferior maxillary bone; 19, incisor teeth; 20, canine teeth; 21, molar teeth.

In the bear, the anterior margin of the frontal bone is prolonged by two small tongues of bone, which, descending on the lateral borders of the nasal bones, articulate with the superior half of the latter.

The temporal bone is, as in man, furnished with a squamous portion, from which springs the zygomatic process, which is directed towards the face, to terminate in the following manner: in the carnivora, the pig, and ruminants, it articulates with the malar bone by its inferior border; in the horse, it insinuates itself as a sort of wedge between the malar bone and the orbital process of the frontal bone, with which it articulates, as we have already pointed out, and contributes, by a portion situated in front of this articulation, to form the boundary of the anterior opening of the corresponding orbital cavity. As in man, the zygomatic process arises by two roots: one, transverse, behind which is situated the glenoid cavity of the temporal bone; the other, antero-posterior, which proceeds to join above with the superior curved line of the occipital bone.

Behind the glenoid cavity is found the external auditory canal, and, further back still, the mastoid process. This latter, but slightly developed in the carnivora, a little more so in the ruminants, and still more in the horse, has its external surface traversed by a crest, *the mastoid crest*, which, after becoming blended with the antero-posterior root of the zygomatic process, proceeds with this latter to join the superior occipital curved line.

Below the auditory canal is situated a round prominence, highly developed in carnivora; this is *the tympanic bulla*, also called *the mastoid protuberance*; it is an appendage of the tympanum.

The Face

The bone of this region, around which all the others come to be grouped, is, as in man, the superior maxillary. The relations of this maxillary with the neighbouring bones is not exactly the same in all animals; for example, in the ox, sheep, and horse, in which the bones of the nose are wide in their upper part, and in which the lachrymal bone, which is very highly developed, encroaches on the face, the superior maxillary does not meet the frontal bone; it is separated from it by the above-named bones. It unites with it, on the other hand, in the dog and the cat. In the bear, it is separated from the bones of the nose by a small tongue of bone which springs from the anterior border of the frontal—a process which we have noticed in connection with this latter.

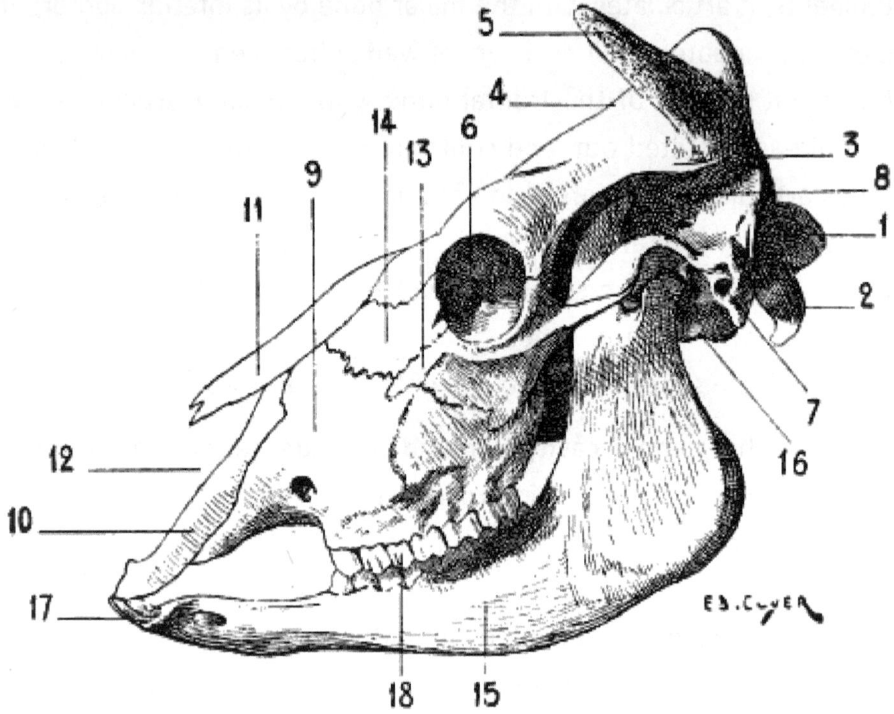

Fig. 62.—The Skull of the Ox: Left Lateral Aspect.

1, Occipital condyle; 2, jugular process; 3, parietal bone; 4, frontal bone; 5, osseous process, which serves to support the horn (horn-core); 6, orbital cavity; 7, external auditory canal, in front of which is found the zygomatic process; 8, temporal fossa; 9, superior maxillary bone; 10, intermaxillary or incisor bone; 11, nasal bone; 12, anterior orifice of the cavity of the nasal fossæ; 13, malar bone; 14, lachrymal bone; 15, inferior maxillary bone; 16, condyle of the inferior maxillary bone; 17, incisor teeth; 18, molar teeth.

In the pig, ox, sheep, and horse, the external surface is traversed, to a greater or less extent, by a crest which is situated on the prolongation of the inferior border of the malar bone. This crest, which is straight in the horse, but curved with its convexity upwards in the ox and the sheep, is known as *the maxillary spine* or *the malar tuberosity*: it gives attachment to the masseter muscle, and, in the horse, is distinctly visible under the skin. It does not exist in the carnivora. On the same surface is situated the sub-orbital foramen.

The inferior border is hollowed out into alveoli, in which are implanted the superior molar and canine teeth. This border is prolonged forwards from the alveolus, which corresponds to the first molar tooth, to terminate, after a course more or less prolonged, at the alveolus of the canine. This space, more or less considerably expanded, which thus separates these teeth is called the *interdental space*; but this denomination is not applicable to ruminants, because these latter possess neither canine nor incisor teeth in the upper jaw (see, dentition of the ox and sheep). The superior maxillary bone of one side and that of the opposite side do not meet in the median line in the region which corresponds to the incisor teeth; they are separated by a bone which, in the human species, is present only at the commencement of life, and afterwards coalesces with the maxilla; this is the intermaxillary or incisor bone. This bone, which is paired, is formed of a central part, which bears the superior incisor teeth; it is prolonged upwards and backwards by two processes: one, external, which insinuates itself between the superior maxillary and the nasal bone, except in the sheep, in which it remains widely separated from the latter; the other, internal, which is united to that which belongs to the bone of the opposite side to form part of the floor of the cavity of the nasal fossæ; the external border of this process, which is separated from the body of the bone by a notch, forms the internal boundary of the corresponding *incisor opening* or the *incisor slit*. Owing to the absence of superior incisors in ruminants, the intermaxillary bone presents no alveoli.

The malar bone, and the os unguis or lachrymal, are more or less developed according to the species considered. With regard to the malar bone, it is most important to notice the part which it takes in the formation of the zygomatic arch, and that its inferior border contributes to form the crest to which is attached the masseter muscle.

As for the nasal bones, they present differential characters which, as they affect the form of the region which they occupy, are worthy of notice.

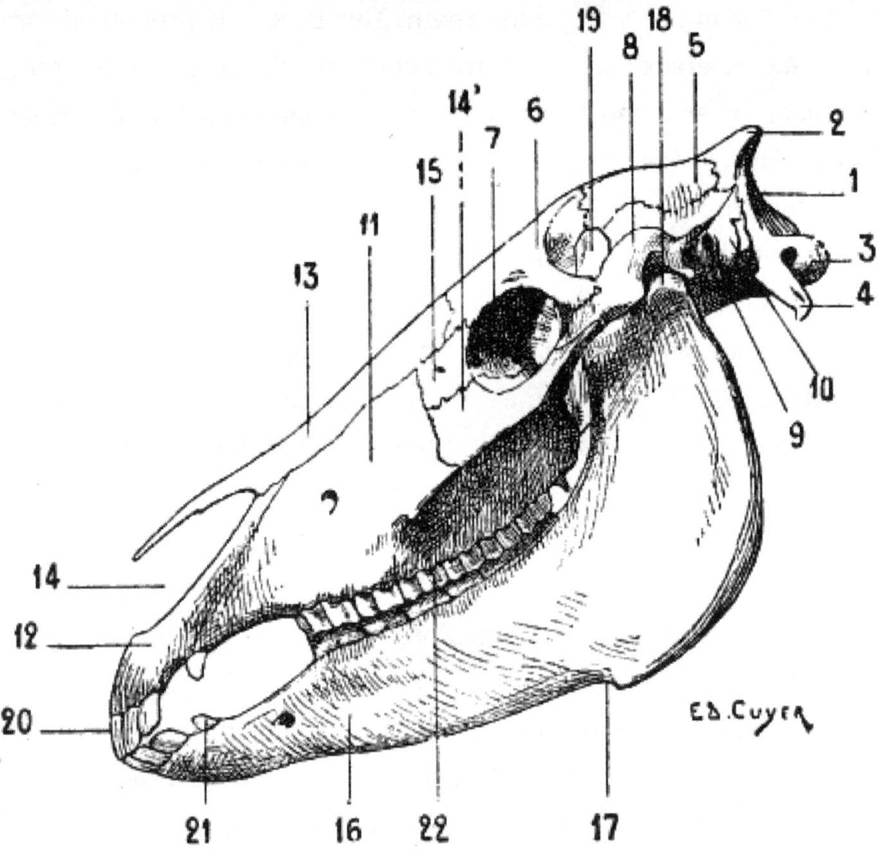

Fig. 63.—Skull of the Horse: Left Lateral Aspect.

1, Posterior surface of the occipital bone; 2, external occipital protuberance; 3, occipital condyle; 4, jugular process; 5, parietal bone; 6, frontal bone; 7, orbital cavity; 8, zygomatic process of the temporal bone; 9, external auditory canal; 10, mastoid process; 11, superior maxillary bone or *maxilla*; 12, intermaxillary or incisor bone; 13, nasal bone; 14, malar bone; 15, lachrymal bone; 16, inferior maxillary bone or *mandible*; 17, inferior maxillary fissure; 18, condyle of the inferior maxillary bone; 19, coronoid process of the inferior maxillary bone; 20, incisor teeth; 21, canine teeth; 22, molar teeth.

Their dimensions in length are proportional to those of the face. Very small in man, they are more developed in carnivora. We recognise in the latter the two curves which characterize them in the human species, and which we clearly notice when we view them on one of their lateral aspects: a concavity above, and a convexity below. These curves are more or less accentuated—very strongly marked in the bulldog, and scarcely at all in the greyhound. Moreover, in the carnivora also the nasal bones are wider below than above, and form, by their junction, a semicircular notch which limits, in its superior portion, the anterior opening of the cavity of the nasal fossæ. In the horse they present an opposite arrangement with regard to their dimensions in width; broad above, each terminates below by forming a pointed process which, separated from the intermaxillary bones, is prolonged in front of the nasal orifice.

The inferior maxillary bone is, as in man, formed of a body and two branches. But among the many special characteristics of form and size which sharply differentiate it from the human bone, one detail must be indicated; this is the absence of a mental prominence. Hence it results that the anterior border of the body of the lower jaw, instead of being directed obliquely downwards and forwards, is, on the contrary, oblique downwards and backwards, and that in certain animals this border is actually found almost exactly on the prolongation of the inferior border of the body of the bone.

On the external surface of the body are found the three mental foramina. The superior border is hollowed out by alveoli.

With regard to the branches (*rami*), they terminate in two processes: one, the posterior, is the condyle; the other, situated more forwards, is the coronoid process, which gives insertion to the temporal muscle. These two processes are separated by the sigmoid notch.

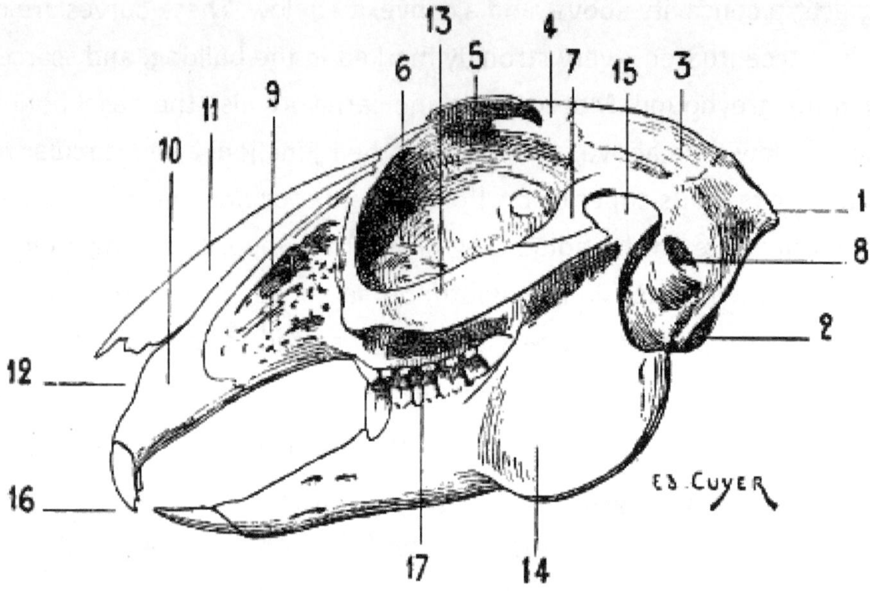

Fig. 64.—Skull of the Hare: Left Lateral Aspect.

1, External occipital protuberance; 2, occipital condyle; 3, parietal bone; 4, frontal bone; 5, orbital process; 6, orbital cavity; 7, zygomatic process; 8, external auditory canal; 9, superior maxillary bone; 10, intermaxillary or incisor bone; 11, nasal bone; 12, anterior opening of the nasal fossa; 13, malar bone; 14, inferior maxillary bone; 15, condyle of the inferior maxillary bone; 16, incisor teeth; 17, molar teeth.

For reasons which we will explain further on (see, movements of the lower jaw), the condyle presents differences of form. In the carnivora, it is strongly convex from before backwards, expanded transversely, and firmly mortised in the glenoid cavity of the temporal bone; in the ruminants, it is less convex from before backwards, it is more slightly concave in the transverse direction; in the rodents—we give as an example the hare (Fig. 64)—the condyle is still convex from before backwards, but it is flattened from without inwards.

In the animals in which the muscles of mastication are very highly developed, and especially in the carnivora, the osseous regions occupied by these muscles are more extensive and more deep than in the human species. The length of the coronoid process, the depth of the temporal fossa, the extent of the zygomatic arch, the appearance of the external surface of each of the rami of the lower jaw, deeply hollowed out for accommodation of the masseter, and to provide extensive surfaces of insertion for this muscle, are sure proofs furnished by the skeleton of the occasionally enormous development of the muscles of mastication.

In the carnivora, a rather strong process, which is directed backwards, occupies the angle of the inferior maxilla; it is, accordingly, situated below the region of the condyle.

The teeth which the jaws carry vary in number, and even in appearance, according to species; it is useful to note their differences. In order to establish the nature of these latter more effectively, we will first recall the fact that in man the teeth, thirty-two in number, are equally distributed between the jaws, and are divided into incisors, canines, and molars, of which the arrangement is thus formulated:

5*m*. 1*c*. 2*i*. 2*i*. 1*c*. 5*m*.

5*m*. 1*c*. 2*i*. 2*i*. 1*c*. 5*m*. = 32.[19]

[19] *I.e.*, *i*, incisors; *c*, canines; *m*, molars.

We also note that the incisors are edged, the canines are pointed, and that the molars, cubical in shape, have their surface of contact provided with tubercles.

The teeth of the cat are thirty in number; they are thus arranged:

4*m*. 1*c*. 3*i*. 3*i*. 1*c*. 4*m*.

3*m*. 1*c*. 3*i*. 3*i*. 1*c*. 3*m*. = 30.

Those of the dog number forty-two:

6*m*. 1*c*. 3*i*. 3*i*. 1*c*. 6*m*.

7*m*. 1*c*. 3*i*. 3*i*. 1*c*. 7*m*.= 42.

In these animals, the incisors, such as are not damaged by use, are furnished, on the free border of their crown, with three tubercles, of which one, the median, is more developed than those which are situated laterally. We denote these teeth, commencing with those nearest the median line, by the names *central incisors* or *nippers, intermediate* and *corner incisors*. The canines, or *fangs*, are long and conical; they are curved backwards and outwards. The upper canines, which are larger than those of the lower jaw, are separated from the most external of the incisors (*corner*) by an interval in which the canines of the lower jaw are received. The lower canines, on the other hand, are in contact with the neighbouring incisors, and are each separated from the first molar which succeeds them by a wider interval than that which is situated between the corresponding teeth in the upper jaw.

The molars differ essentially from the teeth of the same class in the human species. Their crown terminates in a cutting border bristling with sharp-pointed projections; this formation indicates that these teeth are principally designed for tearing. During the movement of raising the lower jaw, which is so energetic in the carnivora, they act, indeed, in the same manner as the two blades of a pair of scissors. The largest molars are: in the dog, the fourth of the upper jaw, and the fifth in the opposite one; in the cat, the third both above and below.

The pig has forty-four teeth disposed in the following manner:

$$7m.\ 1c.\ 3i.\ 3i.\ 1c.\ 7m.$$

$$7m.\ 1c.\ 3i.\ 3i.\ 1c.\ 7m. = 44.$$

Of the incisors, the nippers and the intermediate ones of the upper jaw have their analogues in those of the horse; in the lower jaw, the corresponding teeth, straight, and directed forward, rather resemble the same incisors in rodents. The corner incisor teeth are much smaller, and are separated from the neighbouring teeth. The canine teeth, also called *tusks* or *tushes*, are greatly developed, especially in the male. The molars increase in size from the first to the last; they are not cutting, as in the carnivora, but they are not flattened and provided with tubercles on their surfaces of contact as in the herbivora.

In the ox and the sheep the teeth are thirty-two in number:

6m. 0c. 0i. 0i. 0c. 6m.

6m. 0c. 4i. 4i. 0c. 6m. = 32.

As we see from this dental formula, the incisors are found only in the lower jaw; they are replaced in the upper jaw by a thick cartilaginous pad on which the inferior incisors find a surface of resistance.

These have their crowns flattened from above downwards, and gradually become thinner from the root to the anterior border, which is edged and slightly convex. These teeth gradually wear away. In proportion to the progress of this wear, on account of the fact that it involves the anterior borders and upper surfaces of the incisor teeth, and that these teeth are narrower towards the root than at the opposite extremity, the intervals which separate them tend to become wider and wider; and when the roots become exposed by the retraction of the gums, they are then separated from one another by a considerable interval. The molars have their grinding surface comparable to that of the horse; they increase in size from the first to the sixth.

The teeth of the horse are forty in number; they are thus distributed:

6m. 1c. 3i. 3i. 1c. 6m.

6m. 1c. 3i. 3i. 1c. 6m.= 40.

As they become worn, these teeth continue to grow, and as, on the one hand, this phenomenon takes place throughout the whole life of the animal, and, on the other hand, the process of wear brings out and makes visible at the surface of friction parts formerly deeper and deeper, and of which the configuration varies at different levels, there result special features which permit the determination of the age of the animal by an examination of its jaws. The incisors are called, commencing with those situated nearest the middle line, *central incisors* or *nippers*, *intermediate* and *corner incisors*. The canines, also designated as the *fangs*, exist only in the male. It is exceptional to find them in the mare, and when they exist in this latter they are less developed than those of the horse. The molars have cuboid crowns; the surface of friction is almost square in the case of the upper molars, and is inclined so as to look inwards; in the case of the inferior ones, it is a little narrowed, and is inclined so as to look outwards. In the upper jaw the external surface of the crown is hollowed by two longitudinal furrows; in the lower jaw the same surface has only one furrow, which at times is but slightly marked.

In the hare the teeth are twenty-eight in number:

6*m*. 0*c*. 2*i*. 2*i*. 0*c*. 6*m*.

5*m*. 0*c*. 1*i*. 1*i*. 0*c*. 5*m*.= 28.

The four incisors of the upper jaw are divided into two groups; one of these is formed by the two principal teeth, the other by two very small incisors which are placed behind the preceding.

Having studied the jaws and examined the arrangement of the teeth, we should say a few words on the movements which the lower jaw is able to execute. In man, these movements are varied in character: the jaw is lowered and raised; it can also be projected forwards and drawn backwards, or carried to the right or left side by lateral movements. Owing to the different modes of nutrition of animals, with which the shape of the teeth is clearly correlated, being more specialized than in the human species, the lower jaw is moved in a fashion less varied and in the direction most suitable for the mastication of the foods which form the aliment of the species considered. Moreover, this is plainly shown in the skeleton by the shape of the condyle of the lower jaw (see, different forms of this condyle). In the carnivora, whose teeth, as we have seen, are all cutting ones, the jaw rises and falls; the food then is, if we consider the two jaws, cut as by the blades of a pair of scissors. In the ruminants, the incisors exist only in the lower jaw, but the molars are thick and well developed; the food is ground by these latter as by millstones, and the movements which favour this action are, above all, the lateral. As for the rodents, in which the incisors are formed for filing down and cutting through hard resisting bodies, their lower jaw moves in the antero-posterior direction, in such a way that the inferior incisors alternately advance and recede beneath those of the upper jaw. The free cutting border of these teeth effectively fulfils the function to which they are destined; their constant wear preserves and revivifies the chisel edge which characterizes them, without leading to their destruction, for the incisors in rodents are of continuous growth.

THE SKULL OF BIRDS

The Skull of Birds (Fig. 65).—If, because it is less important from the artistic point of view, we do not consider it necessary to describe in detail the skull of birds, we yet think it useful to indicate, in their general lines, the peculiarities it presents.

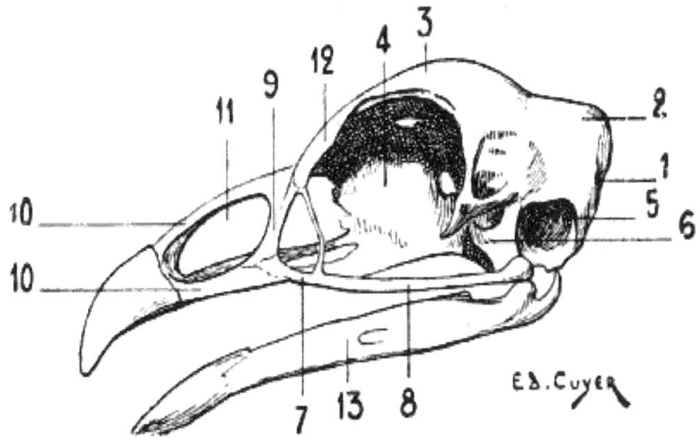

Fig. 65.—Skull of the Cock: Left Lateral Surface.

1, Occipital bone; 2, parietal bone; 3, frontal bone; 4, ethmoid bone; 5, cavity of the tympanum; 6, quadrate bone; 7, superior maxillary bone; 8, malar bone; 9, nasal bone; 10, 10, intermaxillary bone; 11, nasal orifice; 12, os unguis or lachrymal bone; 13, inferior maxillary bone.

In this group the skull is generally pear-shaped; to the cranium, of which the bones are arranged in such a way as to give it a form more or less spherical, succeeds a face more or less elongated, according as the bill is more or less developed.

In general, the bones of the skull coalesce very early, with the result that it is only in very young individuals that we can determine their presence.

We find the skull to consist of an occipital bone, two parietals, a frontal, etc.; we will indicate but one detail in connection with these bones: it is the presence of a single condyle for the articulation of the occipital bone with the atlas. We also note the quadrate bone, which is situated on the lateral part of the cranium, is movable on this latter, and acts as an intermediary between it, the bones of the face, and the lower jaw. The quadrate bone is regarded as a detached portion of the temporal; on the signification of this we do not now propose to dwell.

On the anterior portion of the face we find the nasal bones, which, articulating with the frontal on one side, circumscribe, on the other, the posterior border of the nares. The nasal bone of the one side is separated from that of the opposite by the intermaxillary or premaxillary bone, which forms the skeleton of the superior mandible.

The superior maxillaries, which are rudimentary, are situated on the lateral parts, and prolonged backward by an osseous style which articulates with the quadrate bone; this styloid bone, the homologue of the malar, is designated by certain authors as the *jugal* or *quadrato-jugal* bone.

It is with the quadrate bone also that the inferior maxillary articulates.

CHAPTER II

MYOLOGY

The first point to decide in commencing this study is the order in which we shall consider the different muscles which we have to examine. It must not be forgotten that in the present work we compare the organization of animals with that of man, which we already know, and that it is on the construction of this latter that, in these studies, the thought must at each instant be carried back in order to establish this comparison. Now, the general tendency which we notice in our teaching of anatomy, when one regards the region of the trunk in the human figure (a living model or a figure in the round), is first to consider the anterior aspect. It is the latter that, for this reason, we study at the very beginning; we next deal with the posterior surface of the trunk, because it is opposite; lastly, the lateral surfaces, because they unite with the preceding surfaces, the one to the other.

In studying an animal, it is usually by one of its lateral aspects that one first observes it; it is, in fact, by these aspects that it presents its greatest dimensions, and that the morphological characters as a whole can be more readily appreciated. Hence, possibly, the order of description adopted in most texts, or in the figures which accompany them. The first representation of the human figure as a whole, in a treatise on anatomy, represents the anterior aspect; the first view of the horse as a whole, in a treatise on veterinary anatomy, for example, is, on the other hand, a lateral view.

We break with this latter custom, and, without taking into account the tendency above indicated, we will commence our analysis with the study of the aspect of the trunk, which corresponds to the anterior aspect of the same region in man.

The first muscles usually presented for study to artists being the pectorals, it is their homologues that we will first describe here. We will afterwards describe the abdominal region, then the muscles which occupy the dorsal aspect of the trunk. With regard to the lateral surfaces, they will be found, by this fact alone, almost completely studied, since the muscles of the two preceding (back and abdomen), spreading out, so to speak, over them, contribute to their formation. Nothing further will remain but to incorporate with them the muscles of the shoulder; but these will be studied in connection with the anterior limbs, from which they cannot be separated.

The neck, in man, may be considered in an isolated fashion, because, on account of its narrowness in proportion to the width of the shoulders, it is clearly differentiated from the trunk; for this reason we combine the study of it with that of the head. In animals, because of the absence or slight development of the clavicles, the neck is generally too much confounded with the region of the shoulders to make it legitimate to separate it from that region in too marked a fashion. It will, accordingly, be considered next.

We will then undertake the study of the muscles of the limbs, and end with the myology of the head.

THE MUSCLES OF THE TRUNK

We shall divide them into muscles of the thorax, of the abdomen, and of the back.

Muscles of the Thorax

The Pectoralis Major (Fig. 66, 1, 2; Fig. 67, 3, 4; Fig. 68, 7; Fig. 69, 10; Fig. 70, 11).—Further designated by the name of *superficial pectoral*, this muscle is described in treatises on veterinary anatomy as formed of two portions: an anterior one, called the *sterno-humeral* muscle; the other, situated below and behind the preceding, bearing the name of *sterno-aponeurotic*.

It occupies the region of the breast, and, as a whole, it takes origin from the median portion of the sternum, from which it is directed towards the arm and forearm.

The anterior portion (sterno-humeral muscle)—thick, forming an elevation under the skin, and really constituting the pectoral region—is directed downwards and outwards to be inserted into the anterior margin of the humerus—that is to say, to the ridge which limits in front the spiral groove of this bone.

The other part (sterno-aponeurotic muscle) is situated more posteriorly, and corresponds to the region known in veterinary anatomy as the *inter-fore-limb space*, which is limited laterally on each side by the superior portion of the forearm, of which the point of junction with the trunk bears the name *ars*. Arising from the sternum, as we have above indicated, this portion is directed outwards, to be joined with the terminal aponeurosis of the sterno-humeral, and with that which covers the internal surface of the forearm.

All things considered, the sterno-humeral muscle may be regarded as the representative of the upper fibres of the great pectoral of man, of which the attachments, owing to the more or less complete absence of the clavicle in the domestic mammals, the fibres must be concentrated on the sternum; the sterno-aponeurotic portion then representing the inferior fasciculæ of the same muscle.

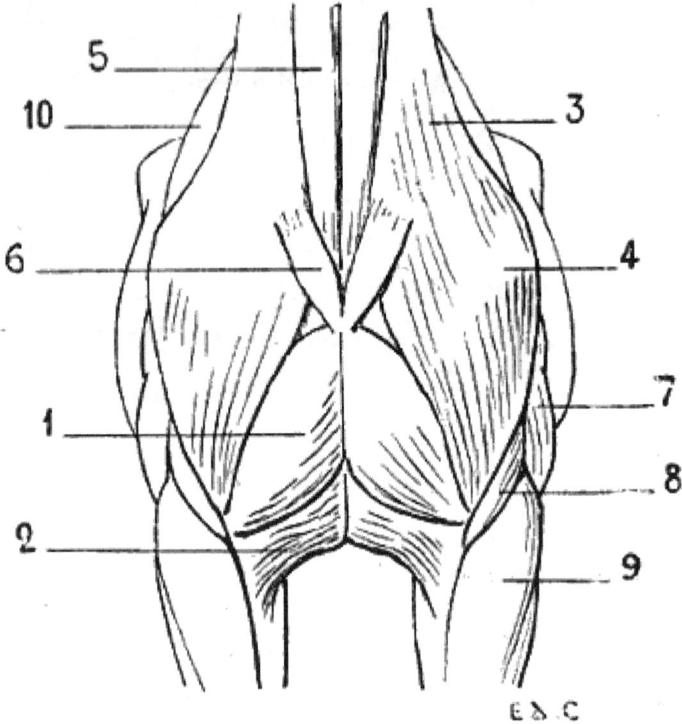

Fig. 66.—Myology of the Horse: Anterior Aspect of the Trunk.

1, Pectoralis major (sterno-humeral); 2, pectoralis major (sterno-aponeurotic); 3, mastoido-humeralis; 4, point of the shoulder; 5, sterno-mastoid or sterno-maxillary: 6, inferior portion of the platysma myoides of the neck, divided; 7, triceps cubiti; 8, brachialis anticus; 9, radialis (anterior extensor of the metacarpus); 10, scapular region.

The great pectoral muscle of one side is separated from that of the opposite side along the median line, and especially above and in front, by a groove which is more or less deep, according as the muscles are more or less developed. At the bottom of this groove, suggestive of that which exists in the corresponding region in man, is found, as in this latter, the median portion of the sternum.

The preceding description particularly applies to the arrangement which the great pectoral presents in the horse; in other animals it is marked by some distinctive characters. In the pig, it is inserted into the sternum as far only as the level of the third costal cartilage; in the ox and sheep, it extends as far as the sixth; in the dog, it is attached to the two first sternal pieces only—that is to say, as far as the third costal cartilage. Moreover, in the latter, as in the cat, the two portions which we have indicated are less readily distinguished.

The great pectoral, by its contraction, draws the fore-limb towards the middle line—that is to say, adducts it.

The Pectoralis Minor (Fig. 67, 6; Fig. 68, 8; Fig. 69, 11; Fig. 70, 12, 26).—This muscle, also called the *deep pectoral*, is, in animals, larger than the superficial pectoral, therefore certain authors prefer to give to this muscle and the preceding one the names of deep and superficial pectoral respectively. This nomenclature is evidently legitimate, and conforms more to reality, since it does not bring in the notion of dimensions which here is found in contradiction to nomenclature; but, in order to establish more clearly the parallelism with the corresponding muscles in man, we think it better, nevertheless, to give them the names by which it has been customary to designate them in connection with the latter.

We will recall at the outset that the lesser pectoral muscle in man is completely covered by the great. In animals this is not the case; the lesser pectoral being very highly developed, projects beyond the great pectoral posteriorly, and occupies to a greater or less extent the inferior surface of the abdomen.

It also consists of two parts: one anterior, which we designate by the name of *sterno-prescapular*; the other, posterior, bearing that of *sterno-humeral*.[20]

[20] This division of the pectorals certainly complicates the nomenclature of these muscles; nevertheless, it introduces no insuperable difficulty from the mnemonic point of view. But where the

study becomes less profitable, and comparison with the corresponding muscles in man more complicated, is in adopting the nomenclature of Bourgelat. Indeed, the great pectoral is designated by this author the 'common muscle of the arm and forearm,' while the lesser pectoral (or deep pectoral) is called the 'great pectoral' in its sterno-trochinian and 'lesser pectoral' in its sterno-prescapular portion. We do not consider it necessary to give the other theories relative to the homologies of these, notwithstanding the very real interest which they present from the purely anatomical point of view, as they have but few applications in the study of forms.

The sterno-prescapular muscle, being covered by the sterno-humeral, has little interest for us. It arises from the sternum, and is directed towards the angle formed by the junction of the scapula and humerus; then it is reflected upwards and backwards, to terminate on the anterior margin of the shoulder by insertion into the aponeurosis, which covers the supraspinatus muscle.

We can, especially in the horse after removal of the skin, recognise it, at the level of this region, in the interspace limited by the superficial muscles (Fig. 70, 26).

In the dog and cat this portion of the muscle does not exist. The other division of the muscle, the sterno-trochinian, is more interesting. It arises from the abdominal aponeurosis and the posterior part of the sternum; hence it passes forward, turns under the superficial pectoral, and is inserted into the lesser tuberosity (trochin) of the humerus.

In the pig, dog, and cat, it is inserted into the greater tuberosity (trochiter) of the bone of the arm.

The superior border of this muscle is in relation with a superficial vein, which is distinctly visible in the horse—the subcutaneous thoracic vein, which in this animal is called the vein of the spur.

The sterno-humeral muscle, in contracting, draws the shoulder and the whole anterior limb backwards.

Serratus Magnus (Fig. 67, 2; Fig. 69, 8; Fig. 70, 9).—This muscle, which is situated on the lateral aspect of the thorax, is covered to a considerable extent by the shoulder, the posterior muscular mass of the arm, and by the great dorsal muscle.

It arises by digitations from the external surface of the dorsal vertebræ; from the first eight in the horse, ox, and dog.

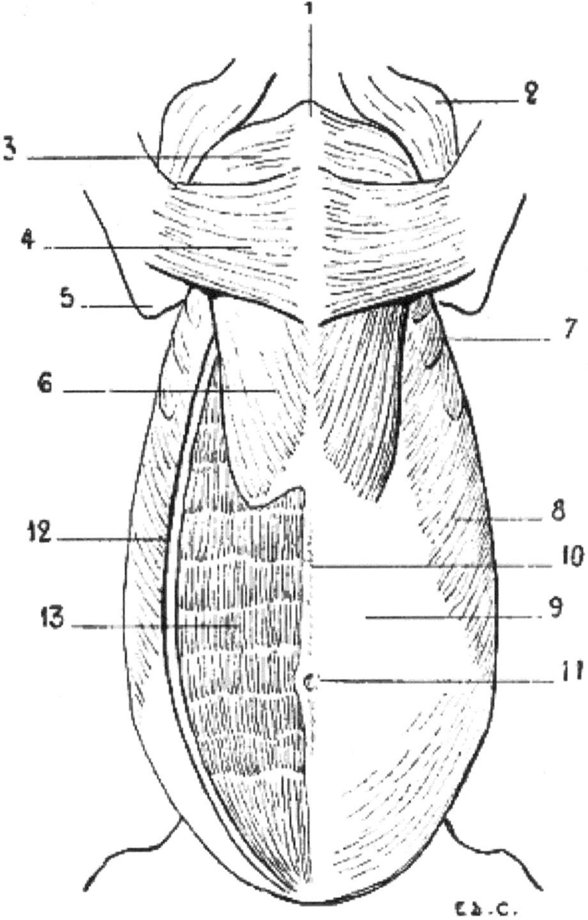

Fig. 67.—Myology of the Horse: Inferior Aspect of the Trunk.

1, Anterior extremity of the sternum; 2, point of the shoulder and inferior portion of the mastoido-humeral muscle; 3, pectoralis major (sterno-humeral); 4, pectoralis major (sterno-aponeurotic); 5, point of the elbow; 6, pectoralis minor (sterno-trochinian); 7, serratus magnus; 8, external oblique; 9, sheath of the rectus abdominis; 10, linea alba; 11, the umbilicus; 12, external oblique divided in order to expose the rectus abdominis; 13, rectus abdominis.

The muscular bundles, converging as they proceed, towards the scapula, pass under this bone, to be inserted into the superior portion of the subscapular fossa, near the spinal border. The inferior portion of its posterior digitations is visible in the ox and in the horse; these digitations are less visible in the pig. They are not seen at all in the dog (Fig. 68) or cat, for in these animals the great dorsal muscle covers them completely.

The great serratus muscle, by the position which it occupies and the arrangement that it presents, forms with the corresponding muscle of the opposite side a sort of girth, which supports the thorax, and at the same time helps to fix the scapula against the latter.

When it contracts, in taking its fixed point at the ribs, it draws the superior portion of the scapula downwards and backwards in such a way that this bone has its inferior angle directed forwards and upwards. If it takes its fixed point at the shoulder, it then acts on the ribs, raises them, and so becomes a muscle of inspiration.

Because of the connections of the serratus magnus with the levator anguli scapulæ, some authors consider it as united with the latter. But as the latter muscle is visible only in the region of the neck, and as it is separately described in man, we prefer to distinguish them from one another. We shall recall the connections to which we have just made allusion when describing the cervical region.

Muscles of the Abdomen

The abdominal wall is, as in man, formed by four large muscles: the external oblique, the internal oblique, and the transversalis, which form the lateral walls, and the rectus abdominis, situated on each side of the middle line of the abdomen. This latter, because of the general direction of the trunk in quadrupeds, has its superficial surface directed downwards.

The arrangement of these muscles closely corresponds to that which we find in the human species.

The External Oblique Muscle (Fig. 67, 8, 12; Fig. 68, 5; Fig. 69, 9; Fig. 70, 10).—This muscle arises, by digitations, from a number of ribs, which varies according to the species, the number of the ribs being itself variable for each of them, as we pointed out in connection with the osteology of the thorax. Indeed, the great oblique arises from the eight or nine posterior ribs in the dog and the ox, and from the thirteen or fourteen posterior in the horse. It is attached, besides, to the dorso-lumbar aponeurosis.

These attachments are arranged in a line directed obliquely upwards and backwards, and the first digitations—that is to say, the most anterior ones—dovetail with the posterior digitations of origin of the great serratus muscle.

The fleshy fibres are directed downwards and backwards, and terminate in an aponeurosis which covers the inferior aspect of the abdomen, and proceeds to form the linea alba by joining with that of the muscle of the opposite side, and also to be inserted into the crural arch.

This aponeurosis of the external oblique is covered by an expansion of elastic fibrous tissue, which doubles it externally, and which is known as the *abdominal tunic*. This latter is further developed as the organs of the digestive apparatus are more voluminous, and their weight, consequently, more considerable. For this reason, in the large herbivora, as the ox and the horse, this tunic is extremely thick, whereas in the pig, cat, and dog it is, on the contrary, reduced to a simple membrane. Indeed, in these latter, the abdominal viscera being less developed, the inferior wall of the abdomen does not require so strong a fibrous apparatus for supporting them. The great oblique, when it contracts, compresses the abdominal viscera in all circumstances under which this compression is necessary; it also acts as a flexor of the vertebral column.

The Internal Oblique Muscle.—This muscle, which is covered by the preceding, arises from the anterior superior iliac spine (external angle in ruminants and solipeds) and the neighbouring parts. From this origin its muscular fibres, the general direction of which is opposite to that of the fibres of the external oblique, diverging, proceed to terminate in an aponeurosis, which contributes to the formation of the *linea alba*, and to be attached superiorly to the internal surface of the last costal cartilages. It has the same action as the great oblique. What it presents of special interest is the detail of form which it determines in the region of the flank; this detail is *the cord of the flank*. It is characterized by an elongated prominence which, starting from the iliac spine, is directed obliquely downwards and forwards, to terminate near the cartilaginous border of the false ribs.

Often very apparent in the ox, and still more so in the cow, the cord in question contrasts with the depression which surmounts it; this depression is situated below the costiform processes of the lumbar vertebræ, and is called the *hollow of the flank*. It is so much the more marked as the mass of the intestinal viscera is of greater weight.

We sometimes meet with a case of the presence of this hollow in the horse. But when in the latter, the flank is well formed, the hollow is scarcely visible, and the cord but slightly prominent. It is only in emaciated subjects that these details are found clearly marked.

Transversalis Abdominis.—This muscle being deeply situated does not present any interest for us. We will, however, point out, in order to complete the series of muscles which form the abdominal wall, that the direction of its fibres is transverse, and that they extend from the internal surface of the cartilages of the false ribs, and the costiform processes of the lumbar vertebræ to the *linea alba*.

The Rectus Abdominis (Fig. 67, 13; Fig. 68, 6).—This muscle, enclosed, as it is in man, in a fibrous sheath (Fig. 67, 9) formed by the aponeuroses of the lateral muscles of the abdomen, is a long and wide fleshy band, which, as in the human species, reaches from the thorax to the pubis.

What distinguishes it in quadrupeds is that there are costal attachments which extend further on the sternal surface of the thorax, and the number of its aponeurotic insertions, which, in general, is more considerable. These are, indeed, six or seven in number in the pig and in ruminants, and about ten in the horse.

It is true that we may find but three in the cat and dog; still, we often find as many as six. These intersections are not marked on their exterior by transverse grooves, such as we find in the human species in individuals with delicate skin and whose adipose tissue is not very much developed.

The rectus abdominis is covered, in its anterior portion, by the sterno-trochinian muscle (posterior segment of the small pectoral). In contracting, this muscle brings the chest nearer the pelvis, and as a result flexes the vertebral column. It also contributes to the compression of the abdominal viscera.

Pyramidalis Abdominis.—This unimportant little muscle, which in man is situated at the lower part of the abdomen, extends from the pubis to the *linea alba*. It is not present in the domestic animals.

We consider it interesting, however, to point out, although the fact is not a very useful one as regards external form, that this muscle is distinctly developed in marsupials.

We know that in the opossum, the kangaroo, and the phalanger fox, the young are brought forth in an entirely incomplete state of development, and that, during a certain period, they are obliged to lodge in a pouch which is placed at the lower part of the abdomen of the mother. Now, this pouch contains the mammary glands; but the young, being too feeble to exercise the requisite suction, the pyramidal muscles come to their assistance. These muscles, in contracting, approximate to one another two bones which are placed above the pubis, the (so-called) marsupial bones (see Fig. 80); by their approximation the bones in question, which are placed behind and on the outer side of the mammary glands, compress the latter, and thus is brought about the result which the little ones, on account of their feebleness, would, without that intervention, be incapable of obtaining for themselves.

Muscles of the Back

Trapezius (Fig. 68, 1, 2; Fig. 69, 1, 2; Fig. 70, 1, 2).—This muscle, more or less well developed, according to the species, is divided into two portions, of which the names indicate the respective situations—a cervical and a dorsal.

These two parts, considered in the order in which we find them, take their origin from the superior cervical ligament and from the spinous processes of the first dorsal vertebræ. From these different points the fibres are directed towards the shoulder; the anterior are, consequently, oblique downwards and backwards, and the posterior are directed downwards and forwards. They are inserted into the scapula in the following manner: the fibres of the dorsal portion are attached to the tuberosity of the spine; those of the cervical region are also fixed into the same spine, but into a considerably larger surface.

The cervical portion occupies, in the region of the neck, an area relatively smaller than the corresponding portion of the trapezius in man. This diminished degree of development results from the absence, complete, or nearly so, of the clavicle in the animals which we are now considering. We remember, that the trapezius of man is partly inserted into the clavicle, and the disappearance of this latter cannot fail to bring modifications in the general disposition of the corresponding portion of the muscle. There results a disconnection of this latter, and it becomes united to other muscular fibres to form a muscle with which we shall soon have to deal—the mastoido-humeral.

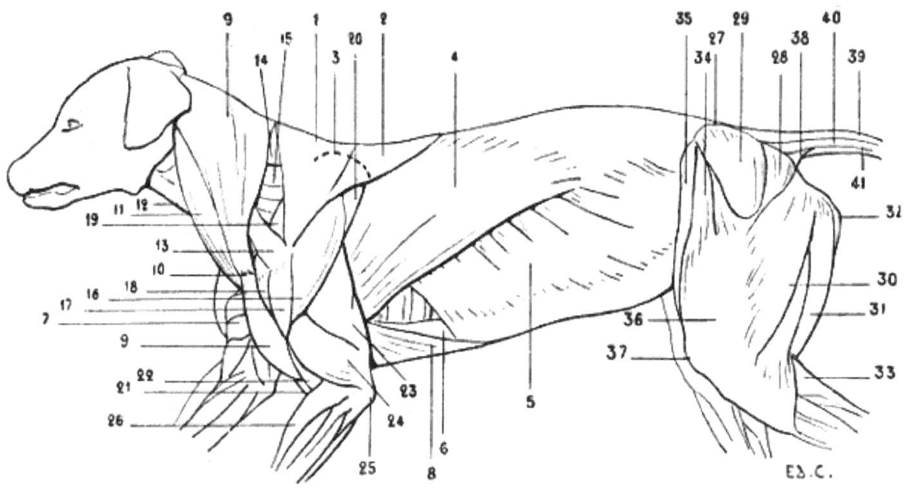

Fig. 68.—Myology of the Dog: Superficial Layer of Muscles.

1, Trapezius, cervical portion; 2, trapezius, dorsal portion; 3, superior outline of the scapula; 4, latissimus dorsi; 5, external oblique muscle; 6, rectus abdominis; 7, pectoralis major of the right side; 8, pectoralis minor (sterno-trochinian); 9, 9, mastoido humeral muscle; 10, tendinous intersection, at the level of which is found a rudimentary clavicle; 11, sterno-mastoid muscle; 12, infrahyoid muscles; 13, omo-tracheal or acromio-tracheal muscle; 14, splenius; 15, levator anguli scapulæ; 16, deltoid muscle, spinal portion; 17, deltoid, acromial portion; 18, superior extremity of the humerus; 19, supraspinatus; 20, infraspinatus; 21, biceps cubiti; 22, brachialis anticus; 23, triceps cubiti, long head; 24, triceps cubiti, external head; 25, olecranon process; 26, radialis (anterior extensor of the metacarpus); 27, iliac crest; 28, gluteus maximus; 29, gluteus medius; 30, biceps cruris; 31, semitendinosus; 32, semi-membranosus; 33, gastrocnemius; 34, tensor of the fascia lata; 35, sartorius; 36, fascia lata drawn up by the triceps; 37, the patella or knee-cap; 38, ischio-coccygeal muscle; 39, superior sacro-coccygeal; 40, lateral sacro-coccygeal; 41, inferior sacro-coccygeal.

As specific differences we should add that the trapezius occupies a more or less extensive portion of the median and superior regions of the neck; terminating at a considerable distance from the head in the dog and horse, it, on the contrary, approaches it in the pig and in ruminants. The cervical portion, when it contracts, draws the scapula upwards and forwards, the dorsal portion draws it upwards and backwards. When the trapezius acts as a whole the scapula is raised.

The Latissimus Dorsi (Fig. 68, 4; Fig. 69, 5; Fig. 70, 5).—This muscle arises by an aponeurosis, the so-called dorso-lumbar aponeurosis, from the spinous processes of the last dorsal vertebræ (the seven last in the dog, fourteen or fifteen last in the horse), from the spinous processes of the lumbar vertebræ, and from the last ribs. Its fleshy fibres are directed downwards and forwards, being more oblique in direction posteriorly, and pass on the inner side of the posterior muscular mass of the arm, to be inserted into the internal lip of the bicipital groove of the humerus, or, a little lower down, on the median portion of the internal surface of the same bone. This latter mode of insertion is met with in the horse and the ox.

The anterior fibres cover the posterior angle of the scapula (as in man, where the corresponding angle, but in this case inferior, is covered by the same muscle), and, a little higher up, are in their turn concealed by a portion of the dorsal fibres of the trapezius. It covers, to a greater or less extent, the great serratus muscle. These relations are similar to those found in the human species.

We find that the fleshy fibres of the great dorsal are prolonged more or less backwards if we examine this muscle in the dog, the ox, the pig, and the horse. Indeed, the fibres reach to the thirteenth rib in the dog and the cat (that is to say, the last rib), the eleventh in the ox, tenth in the pig, and twelfth only in the horse. We say 'only' in connection with this last because it is necessary to remember that the ribs are eighteen in number on each side of the thorax of this animal, and that, accordingly, the fleshy fibres of the great dorsal muscle are, relatively, of small extent.

When this muscle contracts it flexes the humerus upon the scapula, and helps to draw the whole of the anterior limb backwards and upwards.

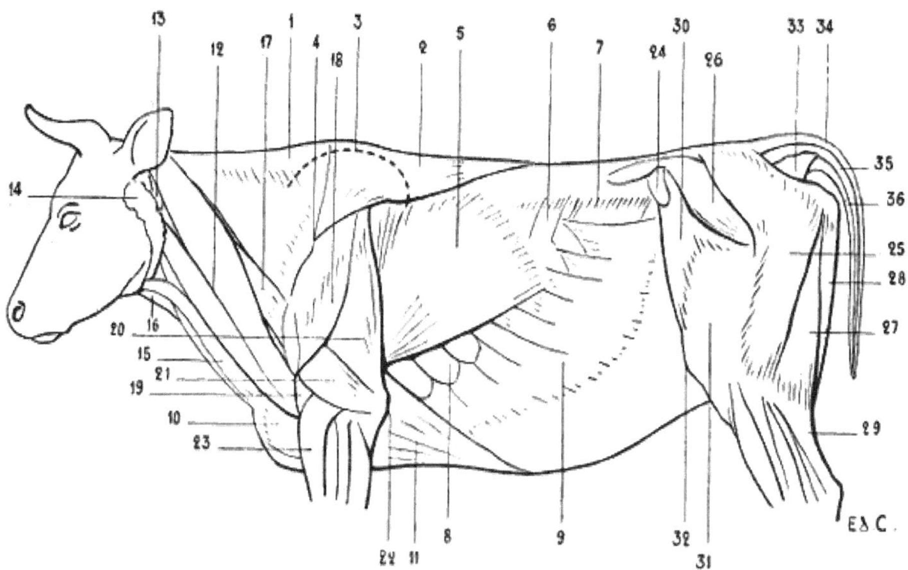

Fig. 69.—Myology of the Ox: Superficial Layer of Muscles.

1, Trapezius, cervical portion; 2, trapezius, dorsal portion; 3, outline of the scapula; 4, spine of the scapula; 5, latissimus dorsi; 6, small posterior serratus; 7, prominence caused by the costiform processes of the lumbar vertebræ; 8, serratus magnus; 9, external oblique; 10, pectoralis major (sterno-humeral); 11, mastoido-humeralis; 12, atlas; 13, atlas; 14, parotid gland; 15, sterno-mastoid muscle; 16, infrahyoid muscles; 17, omo-trachelian or acromio-trachelian muscle; 18, deltoid; 19, brachialis anticus; 20, triceps, long head; 21, triceps, external head; 22, olecranon; 23, radialis (anterior extensor of the metacarpus); 24, anterior iliac spine; 25, gluteus maximus; 26, gluteus medius; 27, biceps cruris; 28, semitendinosus; 29, gastrocnemius; 30, tensor of the fascia lata; 31, fascia lata covering the triceps of the thigh; 32, patella; 33, ischio-coccygeal muscle; 34, superior ischio-coccygeal; 35, lateral ischio-coccygeal; 36, inferior ischio-coccygeal.

There is a muscular fasciculus which, because of its relations with the muscle we have just been studying, is known as the *supplementary muscle of the latissimus dorsi*. But as, on the other hand, this fasciculus is in relation with the triceps, we shall in preference consider it in relation with this latter.

The aponeurosis by which the great dorsal arises from the vertebral column covers, as in man, the muscles which occupy the grooves situated on each side of the spinous processes—the spinal muscles or common muscular mass, if we regard them as a whole (Fig. 70, 7); the sacro-lumbar and the long dorsal muscles covering the transverse spinal, if we consider them as distinct.

It would be superfluous to enter here into a detailed examination of these muscles.

If they are but little developed the spinous processes become prominent under the skin; if they are more so they may by their thickness project beyond the level of these processes, and these latter thus come to lie in a groove more or less marked, which, on account of the division which is determined by its presence, has caused the regions which it occupies to be designated by the names *double back* and *double loins*.

The muscles are extensors of the vertebral column.

Under the aponeurosis of the great dorsal muscle there is found in man another muscle, the serratus posticus inferior, which, on account of being deeply placed and its slight thickness, offers nothing of interest in connection with the study of external form. It arises from the spinous processes of the three last dorsal vertebræ and those of the three first lumbar; it then passes upwards and outwards, and divides into four digitations, to be inserted into the inferior borders of the four last ribs. We repeat that it is covered by the great dorsal muscle.

In the pig, ox, and horse, which have this latter muscle less developed in its posterior portion, the same small serratus muscle, known as the *posterior serratus*, is visible in the superficial layer of muscles (Fig. 69, 6; Fig. 70, 6). The number of its digitations is more or less considerable according to the species examined.

The Rhomboid Muscle (Fig. 70, 21).—In order to make intelligible the position of the rhomboid in the superficial layer in quadrupeds, it appears to us necessary to recall the anatomical characters of the muscle as found in man. The rhomboid arises from the inferior portion of the posterior cervical ligament, from the spinous process of the seventh cervical vertebræ and the four or five upper dorsal; thence passing obliquely downwards and outwards, it is inserted into the spinal border of the scapula, into the portion of this border which is situated below the spine;

it sometimes extends to the middle of the interval which separates this latter from the superior internal angle of the same bone.

The portion of the muscle which arises from the cervical ligament and the seventh cervical vertebra is often separated from the lower portion by a cellular interspace. For this cause some anatomists have described the rhomboid as consisting of two parts—the superior or small rhomboid and the inferior or large rhomboid, on account of the position occupied by each, and of their difference in volume.

This muscle can only be seen in the region of the back, in the space limited externally by the spinal border of the scapula, below by the latissimus dorsi, and internally by the trapezius, which covers it in the rest of its extent. It is not in this space that it is seen in certain quadrupeds. As we pointed out in the section on osteology, the spinal border of the scapula is short, and it seems to be due to this limitation in length that the trapezius and the latissimus dorsi muscle are, at this level, in contact the one with the other in such a way that they fill up the interval in which the rhomboid is seen in man.

In the horse we can partly see it in the superficial muscular layer, but in the region of the neck only, at the superior border of the shoulder. Indeed, as we have already pointed out, the trapezius does not reach the occipital protuberance; for this reason a part of the anterior portion of the rhomboid may be seen—that is, the portion which corresponds to the superior part of the human muscle.

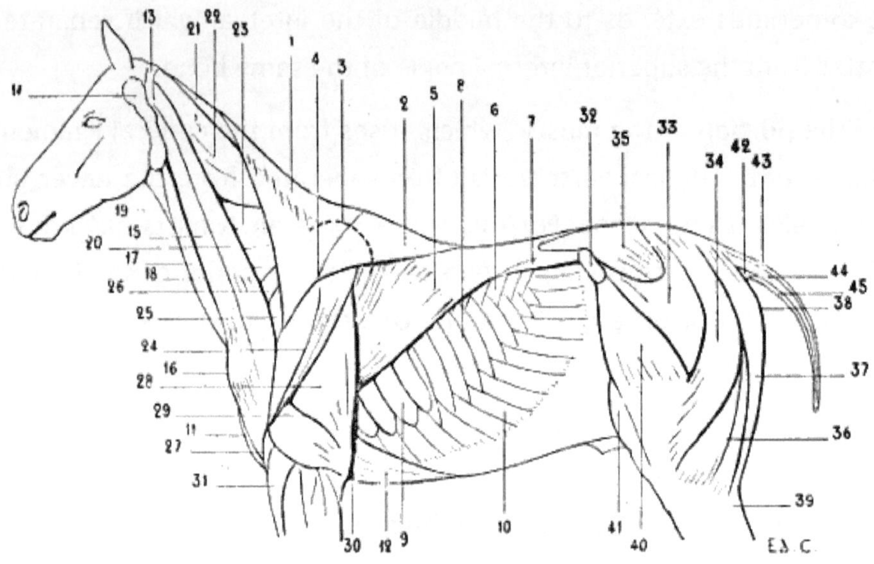

Fig. 70.—Myology of the Horse: Superficial Layer of Muscles.

1, Trapezius, cervical portion; 2, trapezius, dorsal portion; 3, superior outline of the scapula; 4, spine of the scapula; 5, latissimus dorsi muscle; 6, small posterior serratus; 7, spinal muscles, or common muscular mass; 8, ribs; 9, serratus magnus; 10, external oblique; 11, pectoralis major (sterno-humeral); 12, pectoralis minor (sterno-trochinian); 13, atlas; 14, parotid gland; 15, mastoido-humeralis; 16, point of the arm; 17, sterno-mastoid, or sterno-maxillary; 18, jugular groove; 19, infrahyoid muscles; 20, omo-trachelian muscle; 21, rhomboid; 22, splenius; 23, levator anguli scapulæ; 24, deltoid; 25, supraspinatus; 26, terminal part of the sterno-prescapular, a portion of the small pectoral muscle; 27, brachialis anticus; 28, triceps cubiti, middle or long head; 29, triceps cubiti, external head; 30, olecranon; 31, radial extensor (anterior extensor of the metacarpus); 32, anterior iliac spine; 33, anterior portion of the gluteus maximus—the aponeurosis of the muscle has been divided in order to expose the gluteus medius; 34, posterior portion of the gluteus maximus; 35, gluteus medius; 36, biceps cruris; 37, semitendinosus; 38, point of the buttock; 39, gastrocnemius; 40, tensor of the fascia lata; 41, triceps cruris; 42, ischio-coccygeal muscle; 43, superior sacro-coccygeal; 44, lateral sacro-coccygeal; 45, inferior sacro-coccygeal.

But whether it be covered by the trapezius, or, as we find in the cat and dog, by the *mastoido-humeral muscle*, which is very broad in this region, we do not the less recognise its presence; and in the horse and ox, in particular, it forms an elongated prominence beginning at the level of the scapula, and tapering as it ascends, towards the posterior part of the head.

Its origins are similar to those which we have already described in the human rhomboid. It arises from the cervical ligament and the spinous processes of the foremost dorsal vertebræ; its fibres converge and pass to the scapula, to be inserted into its superior or spinal border, or into the internal surface of the cartilage of prolongation.

It assists in keeping the scapula applied to the thoracic cage, and when it contracts, draws the scapula upwards and forwards.

Taking its fixed point at the scapula, it acts on the neck by its anterior fibres, and extends it.

We shall soon have occasion to mention this muscle again, in connection with the study of the muscles of the neck.

The Cutaneous Muscle of the Trunk (Fig. 71).—Immediately beneath the skin which covers the neck, shoulders, and trunk is found a vast cutaneous muscle, analogous to that which, in the human species, exists only in the cervical region.

This thin muscle, whose function is to move the skin which strongly adheres to it, and in this way to remove from it material causes of irritation (insects, for example), is of considerable thickness in the region of the trunk; where it constitutes what certain authors have designated by the name of *panniculus carnosus*. In this region it extends from the posterior border of the shoulder to the thigh, and, in the vertical direction, from the apices of the spinous process of the dorso-lumbar vertebræ to the median line of the abdomen.

Arising above from the supraspinous ligament of the dorso-lumbar and sacral regions (except in the carnivora; see below) by an aponeurosis which, posteriorly, covers the muscles of the hind-limbs, its fibres are directed to the elbow, on which they are arranged in two layers: a superficial, which becomes continuous with the panniculus muscle of the shoulder; and a deep, which passes on the inner side of the shoulder to be inserted into the internal surface of the humerus; this latter exists only in the dog and cat.

The most inferior fibres, behind, at the level of the knee-cap form a triangular process which in the horse receives the name of the *stifle fold*, from the name veterinarians give to the region of the articulation of the knee. This fold of skin, which commences on the antero-internal surface of this region, is directed upwards, and then forwards, to end by gradually disappearing over the corresponding part of the abdomen.

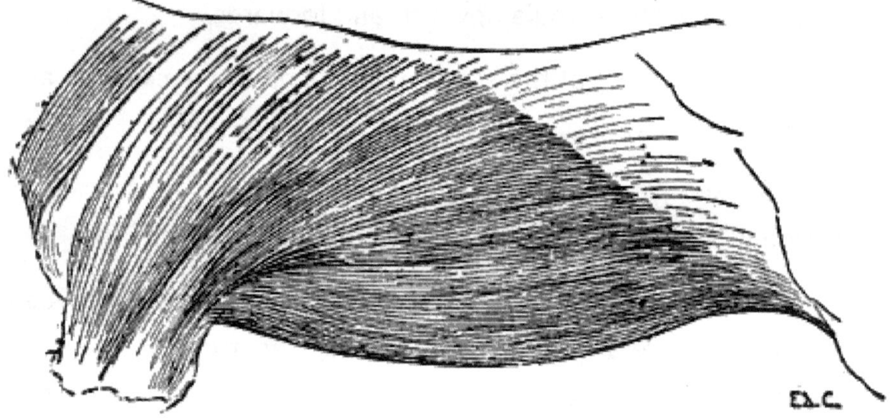

Fig. 71.—Myology of the Horse: Panniculus Muscle of the Trunk.

In the same animal the muscular fibres of the panniculus of the trunk arise along a line which connects the stifle-joint to the withers, a line which is, consequently, oblique upwards and forwards. Now, as the fleshy layer is thicker than the aponeurosis, the result is that the mode of constitution of this muscle can be recognised under the skin. Indeed, we can see in some animals, occasionally very distinctly, a slight elevation starting from the region of the abdomen in the neighbourhood of the knee, and thence directed obliquely upwards and forwards. This elevation is produced by the fleshy portion of the panniculus.

In the carnivora, the panniculus of the trunk is not attached to the supraspinous ligament; it is blended with the same muscle of the opposite side, passing over the spinous region of the vertebral column.

From this arrangement results a great mobility of the skin which covers the back. Further, it explains why it is possible to lift up this skin along with the panniculus which it covers, and to which it adheres, throughout the whole extent of the dorso-lumbar column. As we pointed out above, there is also a panniculus muscle of the shoulder and one of the neck. We will deal with them when treating of the regions to which those muscles belong.

The Coccygeal Region

As a sequel to the study of the muscles of the region of the trunk very naturally comes the description of those which, belonging to the region of the coccyx, are destined for the movements of the caudal appendix, of which this latter constitutes the skeleton. The muscles may not seem to be of much importance with regard to external form, but, as they form part of the superficial muscular layer, and as the mass of each is seen in the form of the tail in some animals (the lion, for example), they merit our attention for a moment. A few lines will suffice to give an idea of them. They are: the *ischio-coccygeal*, *superior sacro-coccygeal*, *lateral sacro-coccygeal*, and *inferior sacro-coccygeal*.

The Ischio-coccygeal (Fig. 18, 38; Fig. 69, 33; Fig. 70, 42).—This muscle, triangular in shape, better developed in the carnivora than in the horse, arises from the spine of the ischium, or from the supracotyloid crest, which replaces this latter in the solipeds and the ruminants. Thence its fleshy mass is directed upwards, expanding as it proceeds to be inserted into the transverse processes of the first two coccygeal vertebræ after insinuating itself between two of the following muscles, the lateral and inferior sacro-coccygeal.

In the dog and cat, the muscle is in great part covered by the great gluteal. In the ox, by a peculiar arrangement of the corresponding region of the muscles of the thigh—an arrangement which we will examine in connection with the study of the latter—it is more exposed than in the horse, and gives origin to an outline which corresponds to its general form in the region situated immediately below the root of the tail.

It is a depressor of the whole caudal appendix.

The Superior Sacro-coccygeal (Fig. 68, 39; Fig. 69, 34; Fig. 70, 43).—The fasciculi which form this muscle arise from the crest of the sacrum, and proceed thence to end successively on the coccygeal vertebræ. It is in contact in the middle line with the corresponding muscle of the opposite side.

It raises the tail and inclines it laterally; if the muscle of one side contracts at the same time as that of the other the tail is elevated directly.

The Lateral Sacro-coccygeal (Fig. 68, 40; Fig. 69, 35; Fig. 70, 44).—Situated on the lateral part of the caudal region, this muscle arises, in the dog, from the internal border of the iliac bone and the external border of the sacrum; in the horse, it arises from the crest of the sacrum. It is inserted into the coccygeal vertebræ.

It produces lateral movement of the tail.

The Inferior Sacro-coccygeal (Fig. 68, 41; Fig. 69, 36; Fig. 70, 43).—This muscle, which is fairly thick, arises from the inferior surface of the sacrum and the corresponding surface of the sacro-sciatic ligament; it is inserted into the coccygeal vertebræ.

It depresses the caudal appendix.

Muscles of the Neck

Mastoido-humeralis (Fig. 66, 3; Fig. 68, 9, 9, 10; Fig. 69, 12; Fig. 70, 15).—One of the most important muscles of the region of the neck in man is the sterno-cleido mastoid. We recollect that, in its inferior part, it is divided into two bundles, one of which arises from the manubrium of the sternum, and the other from the inner third of the clavicle, whence the denominations of the *sternal* portion and *clavicular* portion. The muscle formed by the union of these two portions is then directed obliquely outwards, backwards, and upwards, to be inserted into the mastoid process of the temporal bone and the two external thirds of the superior curved line of the occipital bone.

Now, the animals which we are here considering have but a rudimentary clavicle or are entirely without it. From the absence of this item of the skeleton there necessarily result modifications in the arrangement of the muscles of this region, which we must at the very outset explain, before undertaking the special study of the muscle which is the subject of the present paragraph.

Let us suppose, for the more definite arrangement of our ideas, that the clavicle is altogether absent, although we do find it in a rudimentary state in some animals and completely developed in others (marmot, bat), and we will proceed to indicate what this absence determines.

The great pectoral muscle in man arises in part from the clavicle; this origin not being possible in animals which have no clavicle, its attachments, as we have already seen, are concentrated on the sternum. The trapezius in man similarly arises in part from the clavicle; for the reasons above indicated its clavicular fasciculi cannot exist in distinct form in the animals which have no clavicle.

The sterno-cleido mastoid, whose inferior attachments we mentioned above, cannot have a clavicular portion.

It is the same in the case of the deltoid, which, we know, arises in part from the anterior bone of the shoulder.

Of the four muscles which have partial clavicular origins in man, two are known to us in connection with animals—the great pectoral and the trapezius. What has become of the other two, the sterno-cleido mastoid and the deltoid?

It is this which we now proceed to investigate. After a fashion simple enough, but which it is necessary to describe, the clavicular fasciculi of the trapezius and the corresponding fasciculi of the sterno-cleido mastoid are united the one to the other; the portion of the deltoid which in man arises from the clavicle, by reason of the absence of this latter, is also combined with the fleshy mass formed by the preceding muscles. From this fusion results the muscle known as the mastoido-humeral. This muscle, which consists of a long fleshy band situated on the lateral aspect of the neck, takes its origin, as a general rule, from the posterior surface of the skull and the upper part of the neck, from which it passes obliquely downwards and backwards, covering the scapulo-humeral angle—that is, the region known as the point of the shoulder or arm—and is inserted into the anterior border of the humerus, the border which, limiting anteriorly the musculo-spiral groove, forms a continuation of the deltoid impression. On account of the regions with which it is related, Bourgelat named this muscle *the muscle common to the head, neck, and arm*.

It is at the level of the scapulo-humeral angle that the vestiges of the clavicle are found.

This bone is represented in some animals—the pig, ox, and horse—by a single tendinous intersection, more or less apparent, which extends transversely from the scapula to the anterior extremity of the sternum. In the dog and the cat, we find, besides, on the deep surface of the muscle and at the level of this tendinous intersection, the rudiment of the clavicle of which we made mention in the section on Osteology.

It is beneath the intersection, the existence of which we have just pointed out, that is found that portion of the mastoido-humeral muscle which corresponds to the clavicular fasciculi of the deltoid; that portion which is situated above the intersection corresponds to the clavicular fibres of the sterno-cleido-mastoid and of the trapezius.

The mastoido-humeral presents certain varieties in different animals.

In the dog and the cat, this muscle, which is blended above with the sterno-mastoid, to be inserted with it into the mastoid process and the mastoid crest, covers the neck for a considerable extent from the superior curved line of the occipital bone to which it is attached, to the trapezius with which it unites posteriorly, but from which it separates below. Between these two extreme points of its superior portion it is attached to the cervical ligament.

In the pig and in ruminants, in which the trapezius approaches more closely to the head, the mastoido-humeral occupies, in consequence, a less extent of the cervical region.

In the horse, the mastoido-humeral neither covers the neck nor joins the trapezius; indeed, we have already shown that it is separated by a considerable distance from the head. In the limited interval between these two muscles a part of the rhomboid and parts of other muscles are seen with which we shall soon be occupied.

This muscle, as regards the horse, is described by some anatomists as consisting of two parts: one anterior, or superficial; the other posterior, or deep. In reality, the first only corresponds to the mastoido-humeral, which we are considering; the posterior may be more exactly regarded as representing a special muscle of quadrupeds, but which is here a little deformed, the *omo-trachelian*.

When the mastoido-humeral contracts, taking its fixed point above, it acts as an extensor of the humerus, and carries the entire fore-limb forwards. If it takes its fixed point below—that is to say, at the humerus—it inclines the head and neck to its own side. If it contracts at the same time as the mastoido-humeral of the opposite side, then the head and the neck are carried into the position of extension.

The Sterno-mastoid (Fig. 66, 5; Fig. 68, 11; Fig. 69, 15; Fig. 70, 17).—Having described the clavicular portion of the sterno-cleido-mastoid in connection with the mastoido-humeral, because it forms a part of the latter, we have, in order to complete the homologies of this muscle, to study now that which corresponds to its sternal portion. This is the *sterno-mastoid* muscle. In all the quadrupeds with which we are here concerned this muscle arises from the anterior extremity of the sternum; narrow and elongated in form, it passes towards the head in a direction parallel to the anterior border of the mastoido-humeral, from which it is separated by an interspace which, along its whole length, lodges superficially the jugular vein; hence the name of *jugular groove*, which is given to this part of the neck (Fig. 10, 18).

It is inserted, in the case of the dog and cat, into the mastoid process, where it is united with the mastoido-humeral; in the ox it is divided into two portions—one which goes to the base of the occipital bone, the other passing in front of the masseter is by the medium of the aponeurosis of this latter attached to the zygomatic crest. This latter part is considered by some writers as forming a portion of the panniculus muscle of the neck.

In the horse it is attached to the angle of the lower jaw by a tendon, which an aponeurosis that passes under the parotid gland binds to the mastoido-humeral muscle and the mastoid process.

By reason of this insertion into the jaw, in the case of the solipeds, this muscle is further named the *sterno-maxillary*.

When it contracts, it flexes the head, and inclines it laterally. This movement is changed to direct flexion when the two sterno-mastoid muscles contract simultaneously.

In man, the sterno-cleido-mastoid and the trapezius leave a triangular space between them, which, being limited inferiorly by the middle third of the clavicle, is known as the supraclavicular region; this region, being depressed, especially in its inferior part, has also been given the name of supraclavicular fossa—popularly called the *'salt-cellar.'*

The muscles which form the floor of this region, passing from above downwards, are: a very small portion of the complexus, splenius, levator anguli scapulæ, posterior scalenus, and anterior scalenus; then, crossing these latter, and most superficial, is the omo-hyoid muscle.

An analogous region, but of only slight depth, exists in quadrupeds; its borders are formed by the mastoido-humeral and trapezius muscles.

It is not limited below by the clavicle—we know, indeed, that this, or the intersection which represents it, belongs to the mastoido-humeral muscle—but by the inferior portion of the spine of the scapula.

It is of greater or less extent according to the species considered.

In the dog, cat, pig, and ox, it is narrow, for the muscles which bound it approach one another pretty closely. It has, as in man, the form of a triangle, with the apex above. In the horse it is much broader, and, contrary to the arrangement which it presents in the human species, the widest part is directed upwards.

The muscles which we find there are, consequently, more or less numerous. In the dog and cat they are: a portion of a muscle which we do not normally meet with in man—the *omo-trachelian*—then in a decreasing extent: supraspinatus, levator anguli scapulæ and splenius.

In the pig: the omo-trachelian, supraspinatus, and the terminal portion of the sterno-prescapular—the anterior part of the lesser or deep pectoral muscle.

In the ox: the omo-trachelian only.

But in the horse we find the omo-trachelian, the supraspinatus, and the terminal extremity of the sterno-prescapular; then in a larger extent of area the levator anguli scapulæ and the splenius; and, finally, the anterior portion of the rhomboid.

Among the muscles which we have just enumerated are some that we have already studied; these are the sterno-prescapular and the rhomboid. We will examine the supraspinatus muscle in connection with the region of the shoulder.

As to the scaleni muscles and the complexus, they are deeply situated, whereas the omo-hyoid is visible in the anterior region of the neck only.

There remain for us, accordingly, to examine, at the present juncture, but the omo-trachelian, levator anguli scapulæ, and splenius muscles.

The Omo-trachelian Muscle (Fig. 68, 13; Fig. 69, 17; Fig. 70, 20).—Also called the *acromio-trachelian, levator ventri scapulæ,*[21] the *angulo-ventral muscle,* and the *transverso-scapular,*[22] etc., this muscle is described by some hippotomists as belonging to the mastoido-humeral, of which it then forms its posterior or deep portion.

[21] Ventri, because inserted into the inferior part of the spine of the scapula, towards the acromion—that is, on the ventral side—by contrast with the trapezius, which is attached higher up (dorsal side) on the same process.

[22] Among the many names given to this muscle, Arloing and Lesbre recommend the adoption of the name 'transverse scapular' given by Straus-Durckheim, or 'transverse of the shoulder' (Arloing and Lesbre, 'Suggestions for the Reform of Veterinarian Muscular Nomenclature,' Lyons, 1898).

The omo-trachelian muscle is found in all mammalia, man alone excepted. It is, however, sometimes found in the human being; but it then constitutes an anomaly.

In the dog, pig, and ox, it arises from the inferior part of the spine of the scapula, in the region of the acromion, and terminates on the lateral portion of the atlas.

In the cat it is attached besides to the base of the occipital bone. It is visible in the space limited by the trapezius and the mastoido-humeral, the direction of which it crosses obliquely.

In the horse it appears to be blended in clearly defined fashion with the mastoido-humeral. Attached below, like this latter, to the anterior border of the humerus, it covers the scapulo-humeral angle; and is attached by its upper portion to the transverse processes of the first four cervical vertebræ.

We remember that the transverse processes are often, from their relation with the trachea, known as the tracheal processes. Hence the word 'trachelian,' which forms part of the name of the muscle with which we are now dealing.

By its contraction it helps to draw the anterior limb forwards.

When this muscle, as an abnormality, exists in man, it arises from the clavicle or the acromion process, traverses the supraclavicular fossa, and is inserted into the transverse processes of the atlas or axis, or of both these vertebræ, or of the cervical vertebræ below these latter. It is then known by the names of the *elevator of the clavicle* or *elevator of the scapula*, and, finally, as the *cleido-omo-transversalis* (Testut).[23]

[23] L. Testut, 'Les anomalies musculaires chez l'homme expliquées par l'anatomie comparée,' Paris, 1884, p. 97. A. F. Le Double, 'Traité des variations du système musculaire de l'homme et de leur signification au point de vue de l'anthropologie zoologique,' Paris, 1897, t. i., p. 235.

The Levator Anguli Scapulæ (Fig. 68, 15; Fig. 70, 23).—As we have pointed out (p. 136), the levator anguli scapulæ, because of its connections with the great serratus, is sometimes described with it. But inasmuch as in human anatomy these two muscles are considered separately, and that, in the superficial layer of muscles, they are seen in different regions—the great serratus in the thoracic, and the levator anguli scapulæ in the cervical—we prefer to study them separately.

We remember that in man this muscle arises from the transverse processes of the upper cervical vertebræ and is inserted into the superior portion of the spinal border of the scapula, into the portion of this border which is situated above the spine; it also contributes to the formation of the floor of the supraclavicular region.

When it contracts, it draws the superior portion of the scapula forwards and upwards, and causes a see-saw movement, for at the same time the inferior angle of the scapula is directed backwards. Taking its fixed point at the shoulder, it directly extends the neck if the muscle of one side acts at the same time as that of the opposite; but if only one muscle contracts it inclines the neck to the corresponding side.

It is to be noticed that during movements a little more active than the ordinary the levator anguli scapulæ, as moreover the other muscles of the neck do, becomes very distinct. We have, indeed, often remarked that, apart from these movements, each time the support of one of the fore-limbs is brought into requisition a brusque contraction of the muscles of this region accompanies it.

This contraction gives the impression that, as on the one hand, each support determines a momentary arrest of progression, a jolt, and on the other hand, the head continues to be projected in the forward direction, the latter should be retained. But it cannot be so except by an effort in the opposite direction—that is to say, by the brusque contraction which we have just pointed out.

Analogous contractions also take place in a man while running at the beginning of each contact of the lower limbs with the ground.

We may add, apropos of this latter, that displacements of the head, sometimes in very pronounced fashion, take place during simple walking, and that every time one of the lower limbs is carried forwards the head is projected in the same direction. These displacements, which we also find take place in the horse in pacing, especially in the region of the neck and head, seem then to have the effect of aiding the progression of the body forwards.

They occur especially in animals when drawing a heavy load, and in individuals whose walking movements are executed with difficulty.

It is necessary to repeat that, in these cases, the individual appears to assist the movement of his body by the impetus which the projection of his head forward determines, in order to add—and it is for this that we have referred to the subject—that during the intervals between each projection the head is carried backwards by a muscular contraction similar to that above discussed.

The Splenius (Fig. 68, 14; Fig. 70, 22).—In man, this muscle is attached in the median line to the inferior half or two-thirds of the posterior cervical ligament, to the spinous processes of the seventh cervical, and four or five upper dorsal vertebræ; it passes obliquely upwards and outwards, becomes visible in the supraclavicular region, passes under the sterno-cleido-mastoid, and proceeds to duplicate the cranial insertions of this latter; and, further, the most external fasciculi of this muscle are inserted into the transverse processes of the atlas and the axis.

These separate superior attachments, and the division of the muscle which results, have caused the splenius to be regarded as formed of two portions: splenius of the head, and splenius of the neck.

In the horse, this muscle, which is of voluminous dimensions, arises from the superior cervical ligament, and the spinous processes of the first four or five dorsal vertebræ; thence it proceeds to be inserted into the mastoid crest, and the transverse processes of the atlas and three or four vertebræ following.

The region occupied superficially by the splenius is remarkable for the prominence which this muscle, with the deeply-seated complexus, which is equally bulky, determines at this level; it is situated above that region of the neck, in which are seen in part the fasciculi of the levator anguli scapulæ. It terminates above and in front in the ridge, which is sometimes very pronounced, which the transverse processes of the atlas make on each side of this part of the neck.

In the dog and the cat, the superior and anterior region of the neck is thick and of rounded form, on account of the development which the splenius presents in those animals; but it is covered by the mastoido-humeral.

This latter relation is also found in the ox, but the splenius in this case is but slightly developed.

When the splenius contracts it extends the head and neck, while inclining them to its own side.

If the splenius of one side contracts at the same time as that of the opposite, the extension takes place in a direct manner—that is to say, without any modifying lateral movement.

Infrahyoid Muscles

Having studied the lateral surfaces of the neck, we must now examine the anterior part of this region. Here, between the two sterno-mastoid muscles, we find a space broader above than below, in which are situated the larynx and the trachea, to the general arrangement of which is due the cylindrical form which this region presents. This space corresponds to that which in the neck of man is limited laterally by the sterno-cleido-mastoid muscles, below by the fourchette of the sternum, and above by the hyoid bone. In animals, as in man, it is called the infrahyoid region.

The hyoid bone in quadrupeds is situated between the two rami or branches of the lower jaw. Owing to this disposition, the region above this bone, instead of having its surface projecting a little beyond the inferior border of the maxillary bone, is depressed. This is especially so in the horse. It is there that we find in this animal the region known as the *trough* (*auge*); the larynx corresponds to that part known as the *gullet*.

The muscles which occupy the infrahyoid region are: the sterno-thyroid, the sterno-hyoid, and the omo-hyoid. There is also a thyro-hyoid, but because of its deep situation and its slight importance it offers no interest from our point of view.

Sterno-thyroid and the Sterno-hyoid Muscles.—These two muscles, long, narrow, and flat, arise from the anterior extremity of the sternum; then, covering the anterior surface of the trachea, they proceed to terminate, the one on the thyroid cartilage, and the other on the hyoid bone. The sterno-hyoid is superficial; it covers the sterno-thyroid, which, however, projects a little on its outer side.

Omo-hyoid.—This muscle does not exist in the dog or cat. It arises, in the horse, from the cervical border of the scapula, where it blends with the aponeurosis that envelops the subscapularis muscle, but in the pig and the ox it arises from the deep surface of the mastoido-humeral muscle. It is directed obliquely upwards and inwards, becoming superficial at the internal border of the sterno-mastoid, and is inserted into the hyoid bone.

The region in which are united the portion of the neck which we have just studied and the neighbouring part of the thorax—that is, the breast—has certainly, in our opinion, a form less expressive than the corresponding region in man.

In the latter, indeed, the fourchette of the sternum, with the hollow which it determines, the heads of the clavicles, and the sterno-cleido-mastoid muscles, by the elevations which they produce, and the trachea, by the situation which it occupies in the inferior part, constitute a whole in which are admirably indicated, not only the forms of the organs which constitute this region, but also the relations which these organs have one with another; and, to a certain extent, their respective functions.

In making an exception in the case of the ox, in which a fold of skin, the *dewlap*, which passes from the neck to the breast, constitutes an element of form which possesses some expressive value; in the horse and in the dog, which possess no sternal fourchette and no heads of clavicles, the bones and the muscles are found nearly on the same plane. This produces a uniformity which is evidently inferior, from an æsthetic point of view, to the modelling of the corresponding region of the human body. Such, at least, is our impression.

Suprahyoid Muscles

As their name indicates, these muscles are found above the hyoid bone; amongst those which should arrest our attention for a moment are the mylo-hyoid and the digastric.

Mylo-hyoid.—This muscle, forming a sort of fleshy sling which contributes in great measure to form the floor of the mouth, is situated between the lateral halves of the inferior maxillary bone. Arising on each side from the internal oblique line of the mandible, its fibres are directed towards the median line, to be inserted posteriorly into the hyoid bone, and, between this bone and the anterior part of the mandible, into a median raphe which unites these latter.

Digastric.—This muscle arises from the styloid process of the occipital bone and from the jugular process; it thence passes downwards and forwards, and terminates variously, in different species. In the ox and the horse it terminates in its anterior portion on the internal surface of the inferior maxillary bone, close to the chin. But in the horse a bundle of fibres is detached from the upper portion of the muscle, to be inserted into the recurved portion of the jaw. It is to this fasciculus that Bourgelat has given the name of '*stylo-maxillary muscle*.'

In the pig, dog, and cat, the digastric differs more from the corresponding muscle in man; it is not, as in the latter, formed of two parts. The anterior portion only exists. This consists of a thick muscular mass, which is inserted into the middle of the internal surface of the lower jaw.

In the dog and cat it is clearly recognisable in the superficial layer of muscles by the long and thick prominence which it produces below the masseter, against the inferior border of the mandible.

By its contraction, it draws the lower jaw downwards and backwards.

Panniculus of the Neck.—This very thin muscle, which cannot be recognised on the exterior, calls for little notice.

We shall merely point out that it duplicates the skin of the cervical region; but as the latter is only slightly adherent to it, the panniculus of this region seems rather destined to maintain in position the muscles which it covers than to displace the cutaneous covering.

We recall the fact that in man, on the contrary, the muscle is very evident at the instant of its contraction, and, for this reason, it presents a very great interest with regard to external modelling, and it plays an important part in the expression of the physiognomy.

MUSCLES OF THE ANTERIOR LIMBS

Muscles of the Shoulder

Deltoid (Fig. 68, 16, 17; Fig. 69, 18; Fig. 70, 24).—This is the first muscle we study in connection with the shoulder in human anatomy. Indeed, its wholly superficial position, and especially the manner in which it is separated from the surrounding muscles, its volume, and its characteristic modelling, give it such an importance that, from the didactic point of view, there is every indication for commencing with this muscle in studying the region to which it belongs. If, in regard to quadrupeds, we also commence with it, it is merely in deference to the spirit of method, and for the sake of symmetry; for it is far from presenting, in the latter, characters so distinctive and so clearly defined.

It is necessary to remark, at the outset, that in quadrupeds, on account of the absence or slight development of the clavicle, the clavicular portion of this muscle is, as we have shown, united to bundles of the same kind belonging to the sterno-cleido-mastoid and trapezius to form the mastoido-humeral. There exists, therefore, in an independent form, the scapular portion only.

It is this latter which, by itself alone, forms the deltoid of quadrupeds, a muscle known, in veterinary anatomy, as *the long abductor of the arm*.

In the dog and the cat it consists of two parts, one of which arises from the spine of the scapula; the other from the acromion process. Thence it passes to the crest of the humerus, which limits the musculo-spiral groove anteriorly, to be attached at a point which is found, as in other quadrupeds, to be the homologue of the human deltoid impression, or deltoid **V**, of the human humerus.

In the ox, in which the acromion process, which is very rudimentary, does not attain the level of the glenoid cavity, the acromion portion is but slightly marked off from that which takes its origin from the spine of the scapula.

Still, in the horse, which is completely deprived of an acromion process, the deltoid muscle is correspondingly divided into two parts, separated from one another by superficial interstices, but of which the arrangement differs from that of the portions above indicated; one part, the posterior, arises above from the superior part of the posterior border, and the postero-superior angle of the scapula (exactly as if, in man, certain fasciculi of the deltoid took their origin from the axillary border and inferior angle of the scapula); the other, anterior, arises from the tuberosity of the spine of the same bone. The two parts, united inferiorly, proceed to be inserted into the deltoid impression or infratrochiterian crest of the humerus.

It is necessary to add that the deltoid is inserted into the humerus, above the insertion of the mastoido-humeral.

This muscle flexes and abducts the humerus, and also rotates it outwards.

With regard to the other muscles of the human shoulder, subscapularis, supraspinatus, infraspinatus, teres minor and teres major, they are also present in quadrupeds, but in a form more elongated, as the scapula has its dimensions more extended from below upwards—that is, from the glenoid cavity towards the superior or spinal border.

Subscapularis.—This muscle occupies the subscapular fossa, from which it takes its origin, leaving free the superior part where the surface is found, to which are attached the serratus magnus and the levator anguli scapulæ. It passes towards the arm, to be inserted into the small tuberosity of the humerus. It is an adductor of the arm.

The subscapularis does not offer any interest from the point of view of external form, for it is completely covered by the scapula.

We speak of it, however, because we mention it in human anatomy, and that it affords us here a new opportunity of bringing into prominence the differences which exist in connection with the mobility of the shoulder.

We remember that in man, when the arm is abducted, and then raised a little above the horizontal, the scapula see-saws, is separated, to a certain extent, from the thoracic cage inferiorly and externally, and that, on the superficial layer of muscles, we are then able to see in the bottom of the armpit, at the level of the deep portion of the posterior wall of the latter, a small part of the subscapularis muscle.

In the animals with which we are here occupied it is not the same; for they are incapable of performing with their fore-limbs a movement analogous to that to which we have just referred, the humerus in their case being retained in contact with the trunk by the muscular masses which surround it.

Supraspinatus (Fig. 68, 19; Fig. 70, 25; Fig. 72, 7).—This muscle, as its name indicates, occupies the supraspinous fossa—that is to say, that which, by reason of the direction of the scapula in quadrupeds, is situated in front of rather than above the spine. It arises from this fossa; and, further, from the external surface of the cartilage which prolongs the scapula upwards in solipeds and ruminants. It projects more or less beyond the supraspinous fossa in front.

After passing downwards towards the humerus, it is inserted into the summit of the great tuberosity or trochiter—that is to say, to a part of this osseous prominence which represents the anterior facet of the great tuberosity of the human humerus, into which, as we know, the corresponding muscle is inserted.

In solipeds and ruminants it is inserted, by a second fasciculus, into the small tuberosity.

In the pig and the horse its anterior border is in relation with the terminal portion of the sterno-prescapular anterior portion of the small or deep pectoral.

The supraspinatus, which in man is completely covered by the trapezius, is partly visible in the superficial layer of the cat, dog, pig, and horse, in the lower part of the space limited by the mastoido-humeral and the trapezius. It is crossed by the scapulo-trachelian.

It is, in the ox, completely covered by these muscles, but its form, notwithstanding this, is easily discerned by the prominence which it produces. When it contracts, the supraspinatus muscle carries the humerus into the position of extension.

Infraspinatus (Fig. 68, 20; Fig. 72, 8).—This muscle, which occupies the infraspinous fossa, which, in quadrupeds, is situated behind the spine of the scapula, arises from the whole extent of this fossa, and in solipeds and ruminants encroaches on the cartilage of prolongation. Its fibres are directed downwards and forwards, to be inserted into the great tuberosity of the humerus—the trochiter—below the insertion of the supraspinatus.

It is completely covered (ox and horse), or in part only (cat and dog), by the portion of the deltoid which arises from the spine of the scapula; nevertheless, its presence is revealed by the prominence which it produces.

It is an abductor and external rotator of the humerus.

In connection with this muscle, which, as we have just pointed out, is less seen in the superficial muscular layer than the supraspinatus, we will draw attention to the fact that this arrangement is exactly the reverse of that which is found in the human shoulder. In this latter it is the supraspinatus which is not visible; while, on the contrary, the infraspinatus is uncovered in a considerable part of its extent. We further notice that it is accompanied by the teres minor, and that the teres major, situated inferiorly, forms with these two muscles a fleshy mass which, below, ends on the superior border of the great dorsal muscle.

In quadrupeds, in which the infraspinatus is so slightly visible, the teres major and minor are not found at all in the superficial muscular layer.

Accordingly, we will say but few words about them.

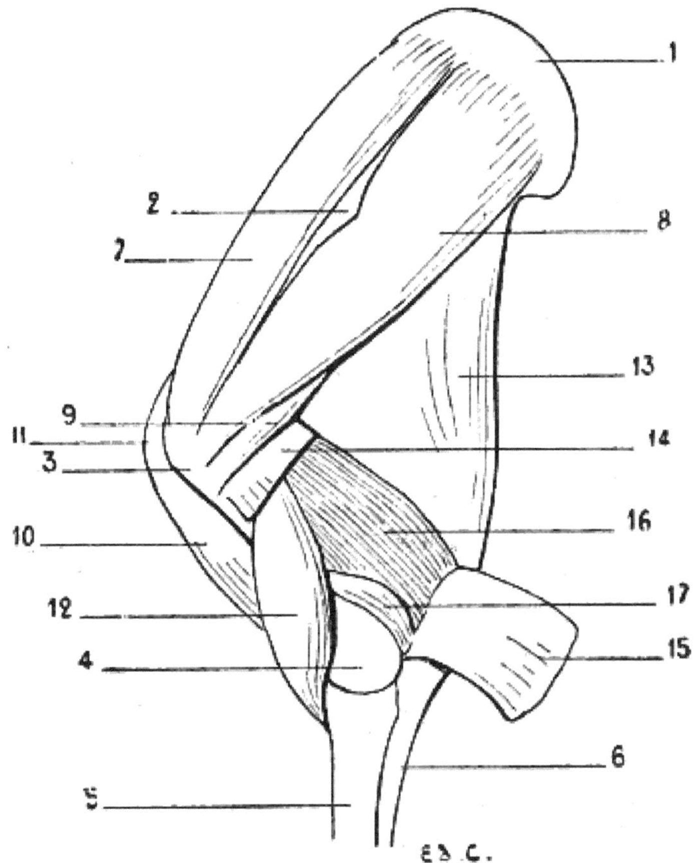

Fig. 72.—Myology of the Horse—Shoulder and Arm: Left Side, External Surface.

1, Cartilage of prolongation of the scapula; 2, tuberosity of the spine of the scapula; 3, superior extremity of the humerus; 4, inferior extremity of the humerus; 5, radius; 6, ulna; 7, supraspinatus muscle; 8, infraspinatus; 9, teres minor; 10, biceps; 11, tendon of the biceps passing over the anterior surface of the superior extremity of the humerus; 12, brachialis anticus; 13, triceps, long head; 14, external head of the triceps divided; 15, external head of the triceps reflected, in order to expose the anconeus; 16, region normally occupied by the external head of the triceps; 17, anconeus.

Teres Minor (Fig. 72, 9).—This muscle, also called in veterinary anatomy *the short abductor of the arm*, arises from the posterior border of the scapula (the external border in man), and is inserted below the great tuberosity of the humerus, between the attachments of the infraspinatus and deltoid.

It is covered by the deltoid and the infraspinatus.

Teres Major.—This muscle is known to veterinarians as *the abductor of the arm*; it arises from the postero-superior angle of the scapula (the inferior angle of the human scapula), from which it passes to be inserted into the internal surface of the humerus.

It is covered by the latissimus dorsi and the posterior muscular mass of the arm.

In brief, for the better understanding of the relations of the teres major and minor muscles in quadrupeds, we may fancy the corresponding muscles in man modified in the following manner: The infraspinatus, thicker, covering the teres minor; latissimus dorsi, more extended in its superior part, covering a large proportion of the teres major. As to the relations of the teres minor with the deltoid, they exist in man, seeing, in this case, the same muscle is, in its external portion, covered by this latter. With regard to the relations of the teres major with the posterior muscular mass of the arm, they also exist in man, since the external surface of this muscle is covered by the triceps.

These modifications are sufficient to render the small and large teres muscles completely invisible in the superficial layer.

The muscles of the shoulder which we have just been studying fulfil, with regard to the articulation which they surround, the function of active ligaments. This rôle is made necessary by the laxity of the scapulo-humeral capsule—a laxity which renders it incapable by itself of maintaining the bones in contact at this joint.

The same condition exists in man.

Panniculus Muscle of the Shoulder.—This thin muscle covers, as its name implies, the region of the shoulder, and is the continuation forward of the panniculus muscle of the trunk.

It arises, by its superior part, from the region of the withers and from the superior cervical ligament; thence its fibres descend directly towards the elbow, to terminate at the level of the region of the forearm.

The muscle is not found in the pig or in the carnivora.

Muscles of the Arm

We should remember, at the outset, that in man the muscles of the arm are divided into two groups: one anterior, which contains the biceps, brachialis anticus, and the coraco-brachialis; the other, posterior, which is constituted by a single muscle, the triceps.

In animals, we find them in the same number and arranged in analogous fashion—that is to say, in two groups—with respect to the bone of the arm. But then we find that they have undergone a transformation with regard to their length, and it is the change of general aspect which results from this modification that we proceed to examine.

We know that in quadrupeds, and especially in the domestic animals, the humerus is relatively short in proportion to the forearm. We have already seen, in dealing with the bones, that whilst in the human species the humerus is longer than the forearm, in the dog and cat these two segments of the fore-limb are of equal length, and that the humerus of the horse is, on the contrary, much shorter. Now, let us suppose the human humerus to be shorter than it is in reality; the anterior muscles undergoing, very naturally, the same reduction, will be uncovered only slightly by those above—the deltoid and the great pectoral—or will remain completely hidden by them. Thus would be found realized the disposition which we meet with in quadrupeds of the muscles of this region.

With regard to the posterior muscular mass of the arm, it does not undergo the same change. The muscle which constitutes it—the triceps cubiti—occupies, on the contrary, a greater area. Let us suppose, further—for it is the best method of comprehending the homologies which now occupy our attention—the humerus of man to be shortened as before, and directed downwards and backwards (as in quadrupeds), this bone would form an acute angle with the axillary border of the scapula. Let us suppose also that the long portion of the triceps, instead of arising solely from the superior part of this axillary border, is attached to the whole length of the latter, and that the triceps fills the whole interior of the angle formed by the arm and the shoulder. We then shall have an idea of what the triceps is in quadrupeds. It is necessary to add that the general resemblance would be still more complete if the arm were firmly supported by the side of the thorax, because in quadrupeds it occupies an analogous position, determined by the arrangement of the muscles which, proceeding from the trunk and neck, are attached to it.

Anterior Region

Biceps Cubiti (Fig. 68, 21; Fig. 72, 10, 11).—This muscle, also called *the long flexor of the forearm*, does not merit the name except by its analogy with the corresponding muscle in man. Indeed, in the domestic animals it is not divided into two parts; it is represented by a single fasciculus, long and fusiform, situated on the front of the humerus, and directed obliquely downwards and backwards, as the latter, on its part, is also inclined.

It arises above from a tubercle at the base of the coracoid process, which surmounts the glenoid cavity of the scapula. Its tendon, which is highly developed in the solipeds, occupies the bicipital groove. We remember that in these latter the groove in question is divided into two channels by a median prominence.

The tendon in which the muscle ends is inserted into a tuberosity, situated on the internal surface of the superior extremity of the radius—the bicipital tuberosity. In the pig, the cat, and the dog, there is detached from the tendon to which we have just referred a fasciculus of the same nature, which, after having wound round the radius, is inserted into the internal surface of the ulna, towards the base of the olecranon process. From the inferior part of the muscle arises a fibrous band, comparable to the aponeurotic expansion of the human biceps; but, instead of passing downwards and inwards, as does the latter, it terminates on the muscular mass which constitutes the antero-external part of the forearm.

The biceps is not seen in the superficial layer, except in the dog and cat (in which the humerus is, in fact, proportionately long); and even in them only to the slightest extent. It is covered partly in these latter, and completely in other animals, by the great pectoral and the inferior portion of the mastoido-humeral—that is to say, that part of the latter which represents the whole of the clavicular fibres of the human deltoid.

The biceps is a flexor of the forearm on the arm. It also contributes to the movement of extension of the humerus.

Brachialis Anticus (Fig. 68, 22; Fig. 69, 19; Fig. 70, 27; Fig. 72, 12).—In veterinary anatomy further designated as *the short flexor of the forearm*, this muscle, which is thick, occupies the musculo-spiral groove, and arises from it, reaching upwards to just below the head of the humerus. But it does not, as in man, extend to the internal surface of the bone.

Situated on the outside of the biceps, it is directed towards the forearm, and terminates by a flattened tendon, which, dividing into two slips, passes below the bicipital tuberosity, on the internal surface of the radius, into which one of these slips is inserted, while the other proceeds to terminate on the ulna.

The inferior half of this muscle is visible on the superficial layer, in the space limited posteriorly by the triceps brachialis, and below by the muscles of the forearm, which correspond to the external muscles of the human forearm, and in front by the great pectoral and the mastoido-humeral. It is in the upper part of the interspace which separates these latter from the brachialis anticus that the deltoid insinuates itself to proceed to its insertion into the humerus.

These relations precisely recall those which we meet with when we examine the external surface of the human arm, with this difference, however—that in the latter the anterior brachialis anticus is extensively related, in front, to the biceps. However, in animals it is not absolutely the same, since, as we have shown above, the biceps is covered, more or less completely, by the mastoido-humeral and the great pectoral.

The brachialis anticus flexes the forearm on the arm.

Coraco-brachialis.—In man this muscle, which occupies the superior half, or third, of the internal surface of the humerus, is visible only when the arm is abducted, and then especially when it approaches the vertical position; indeed, it is only in this attitude that the region which it occupies is accessible to view.

But an analogous attitude not being possible in domestic animals, in which the arm is fixed along the corresponding parts of the trunk, the result is that the coraco-brachialis is always covered, and that, consequently, it presents nothing of interest from our point of view. We speak of it, then, merely in order to complete the series of the muscles of the anterior surface of the arm, among which we rank it, in spite of the fact that in veterinary anatomy it is described as a muscle of the shoulder.

It arises above from the coracoid process, and thence passes downwards towards the internal surface of the humerus into which it is inserted, more or less high up, according to the species. The coraco-brachialis is an adductor of the arm.

Posterior Region

Triceps Cubiti (Fig. 68, 23, 24; Fig. 69, 20, 21; Fig. 70, 28, 29; Fig. 72, 13, 14, 15, 16).—This muscle, which is voluminous in the quadrupeds with which we are here concerned, fits more or less completely the angular space between the scapula and the humerus. Its bulk forms a thick prominence, which surmounts the elbow and the forearm.

We should say, with regard to this mass, that if the deltoid does not constitute in quadrupeds a prominence sufficient to remind one of that which this muscle produces in man, the triceps, in producing an analogous elevation, seems to replace in the general form of the body the relief which the deltoid is incapable of producing.

The triceps is divided into three portions, which, as in man, have the names middle, or long head; external and internal heads. But that which renders the nomenclature a little complicated is that veterinary anatomists have given other names to these three parts: that of *great extensor of the forearm* (caput magnum) to the long head; *the short extensor of the forearm* (caput parvum) to the external head; and of *medium extensor of the forearm* (caput medium) to the internal.[24]

[24] Other names given by certain authors to the parts of this muscle which we have just enumerated still further complicate this nomenclature.

The long head is further designated by them under the names of the *long* or *great anconeus*; the *external head* under those of *external anconeus*, or *lateral* or *short anconeus*; whilst the internal head becomes the *internal anconeus*, or *median*.

It is more especially the long portion and the external head which, being visible on the external surface of the arm, contribute to the external form.

The long portion, which is triangular in shape and of considerable development, arises in the cat and the dog from the inferior half or two-thirds of the posterior border of the scapula (axillary border); from the whole extent of that border as far as the superior posterior angle in the pig, the ox, and the horse; it then passes downwards towards the articulation of the elbow, to terminate in a tendon which is inserted into the olecranon process. The portion of this muscle which is next the scapula is covered by the deltoid.

The external head, situated below the long portion, is directed obliquely downwards and backwards. It arises from the curved crest which, from the deltoid impression of the humerus, is directed upward to meet the articular head of the same bone. This crest limiting the musculo-spiral groove superiorly, and the brachialis anticus arising from the whole extent of this groove, the result is that at this level the external head is in relation with the brachialis anticus. From this origin it is directed towards the elbow, to be inserted into the olecranon, either directly or by the medium of the tendon of the long portion. The part of this muscle which arises from the humerus is covered by the deltoid.

As for the internal head (Fig. 76, 4), which, in the superficial layer, is only visible in its inferior part, on the internal aspect of the arm in those animals in which the elbow is free of the lateral wall of the thorax (the dog and the cat, for example), it arises from the internal surface of the humerus, and thence proceeds to be inserted into the olecranon.

The triceps extends the forearm on the arm.

A fourth muscle exists, which veterinary anatomists include in the study of the three portions of the triceps which we have just been discussing, in giving it the name of *small extensor of the forearm*. But, as this muscle is no other than the anconeus, and as, in human anatomy, we describe the latter, according to custom, in connection with the forearm, it is when on the subject of the latter that we will concern ourselves with it. This grouping of muscles cannot fail to give greater clearness to the description of the muscles of these regions.

The Supplemental or Accessory Muscle of the Latissimus Dorsi (Fig. 76, 2; Fig. 77, 1).—Because of the relations, to which we have already referred, of this muscle with the triceps cubiti, its description very naturally follows that of the latter.

Indeed, this supplementary muscle of the great dorsal is further designated in zoological anatomy under the name of *long extensor of the forearm*; and this name indicates that its study may be united to that of the triceps.

Situated on the internal surface of the arm, it arises from the external aspect of the tendon of the latissimus dorsi; it is very highly developed in the horse, in which it also arises from the posterior border (axillary) of the scapula; then, covering in part the internal head of the triceps and also the long portion, on the superior border of which it is folded, it proceeds to be inserted into the olecranon process and the anti-brachial aponeurosis.

It extends the forearm on the arm. Further, it makes tense the aponeurosis into which it is inserted; this explains the name of *tensor of the fascia of the forearm*, which is sometimes given to it.

It seems to us interesting to add that, abnormally, we sometimes find in man an analogue of this muscle. It is given off from the latissimus dorsi, near the insertion of the latter into the humerus; it accompanies the long head of the triceps and becomes fused with it. Sometimes it is inserted into the olecranon process, at other times into the antibrachial aponeurosis or the epitrochlea. It is on account of its insertion into the last-mentioned, in some cases, that it is also designated by the name of *dorso-epitrochlear* muscle.[25]

[25] L. Testut, 'Anomalies musculaires chez l'homme expliquées par l'anatomie comparée,' Paris, 1884, p. 118. A. F. Le Double, 'Traité des variations du système musculaire de l'homme et de leur signification au point de vue de l'anthropologie zoologique,' Paris, 1897, t. i., p. 203. Édouard Cuyer, 'Anomalies musculaires' (*Bulletins de la Société Anthropologique*, Paris, 1893).

Muscles of the Forearm

Before commencing the special examination of each of the muscles of this region, it is absolutely indispensable to consider their general arrangement, and to determine very clearly how we should study them. We are too well convinced of the importance of this preliminary examination to dismiss it without entering rather fully into it. Indeed, the region on the myological study of which we are now entering is, unquestionably, one of the most complicated with which we have to deal. We know besides, in regard to the study of the forearm in man, how much a definite method is necessary in order that the arrangement of the muscles of this region be fixed in the memory, and that we are unable to obtain this result otherwise than by grouping the twenty muscles which constitute it in clearly defined regions.

We also know that these muscles are first studied with the forearm in the position of supination, and that it is only when they are well known after having considered them in this position that we are able to analyze and comprehend their forms when it is in pronation.

Now, as we have pointed out in the section on osteology, the forearm in quadrupeds is always in the position of pronation. Should we, then, in order to maintain the symmetry with human anatomy, first study the forearm in the position of supination? Evidently not. Besides the fact that this would in some cases be impossible since—as in the horse, for example—the radius and ulna are fused together, we should not gain any advantage; this position being never completely realizable even in those quadrupeds which have the radius relatively movable—as, for example, in the cat.

Accordingly, it is pronation which here, in connection with animals, becomes the standard attitude from the point of view of description. This is why, supposing that the reader knows well the muscles of the human forearm in the position of supination, we should recall what is the general arrangement occupied by these muscles when it is in pronation.

The fore-limb, being viewed on its anterior surface, presents above the anterior aspect of the region of the elbow; but below, it is the posterior surface of the wrist which is seen. Consequently, in the superior part, we see the external and anterior muscles limiting the hollow in front of the elbow; interiorly are found the posterior muscles.

The long supinator, passing obliquely downwards and inwards, divides, in fact, the forearm into two parts: one supero-internal, the other infero-external. In the first we see, but to an extent less and less considerable, the pronator teres, the flexor carpi radialis, the palmaris longus, and the flexor ulnaris; as to the flexors of the digits, on account of the rotation of the radius, they are only visible on the opposite surface—that is to say, on the surface of the wrist, which is now posterior. In the second part we see the two radial extensors, the common extensor of the fingers, the proper extensor of the little finger, and the ulnar extensor which, inferiorly, remains behind, by reason of the position of the ulna being unchanged, whilst the anconeus is wholly posterior, since the direction of the elbow is not modified. We also find, in this region, the long abductor of the thumb, the short extensor of the thumb, the long extensor of the thumb, and the special extensor of the index-finger, in the region where these deep muscles become superficial.

So that, to summarize, the external and posterior muscles occupy the anterior and external regions of the forearm, whilst the anterior muscles occupy rather the internal and posterior. It is in regarding them after this manner—that is to say, arranged in these two regions—that we proceed to study these muscles in quadrupeds.

Anterior and External Region

Supinator Longus.—We know that this muscle, which is especially a flexor of the forearm on the arm, plays, notwithstanding the name which has been given it, a part of but little importance in the movement of supination.

It acts slightly, however, as a supinator, for, being very oblique downwards and inwards at the time of pronation, it is able, while tending to resume its vertical direction, to carry the radius outwards; it places, in fact, the forearm in a position midway between pronation and supination.

We have just recalled these details, in order that it may be more easy to understand why it does not exist in animals in which the radius and ulna are fused together (horse, ox); and why, on the other hand, we find traces of it in the cat and the dog, in which the radius—to a slight extent, it is true—is able to rotate on the ulna. This displacement being a little more considerable in the felide, the long supinator is a little further developed than it is in the canine species; but, notwithstanding, it is only rudimentary.

The long supinator arises, above, from the external border of the humerus; thence, in the form of a very narrow fleshy band, it passes obliquely downwards and inwards, to be inserted into the inferior part of the internal surface of the radius.

It assists in turning the radius outwards and placing it in front of the ulna, the movement of supination being capable of being but little further extended.

First and Second External Radial Muscles: *Extensor carpi radialis longior and brevior* (Fig. 73, 8; Fig. 74, 8, 9; Fig. 75, 8, 9).—Fused together, these muscles form by their union what veterinary anatomists call *the anterior extensor of the metacarpus*. But we should add that these two muscles are united so much the more intimately as we examine them in passing successively from the cat to the dog, pig, ox, and horse. Thus, in the cat they are often distinct; in the dog, they unite only at the level of the middle third of the radius, and interiorly they have two tendons; in the pig, the ox, and the horse they are completely united, and there exists but a single tendon.

The *anterior extensor of the metacarpus*, which is situated behind the long supinator when the latter exists, occupies the external aspect of the forearm; its well-defined form absolutely recalls the prominence on the superior part of the external margin of the human forearm.

It arises superiorly from the portion of the external border of the humerus which is situated above the epicondyle and behind the musculo-spiral groove. Its fleshy mass appears in the angular space bounded by the brachialis anticus and the triceps. The superior portion is covered by the external head of the triceps; yet, in the dog, the superior portion of its humeral attachment is the only part so covered. This muscle is directed forward and downwards; it is also inclined a little inwards in such manner as to proceed to occupy the anterior aspect of the forearm.

Its fleshy belly is narrowed below, and, towards the inferior part of the forearm, is continued by a tendinous portion which is situated on the anterior surface of the carpus, after having traversed the median groove of the inferior extremity of the radius.

In the cat and the dog, in which the union of the two radial extensors is incomplete, the two tendons are inserted into the front of the base of the second and third metacarpal bones; consequently, as in man, into the metacarpals of the index and middle fingers.

In the ox, the tendon, which is single, is inserted into the internal and anterior half of the superior extremity of the principal metacarpal.

In the pig, this tendon is attached to the base of the large internal metacarpal.

In the horse, the corresponding tendon is attached to a tubercle which is situated on the anterior surface of the base of the principal metacarpal, a little internal to the median plane of the latter.

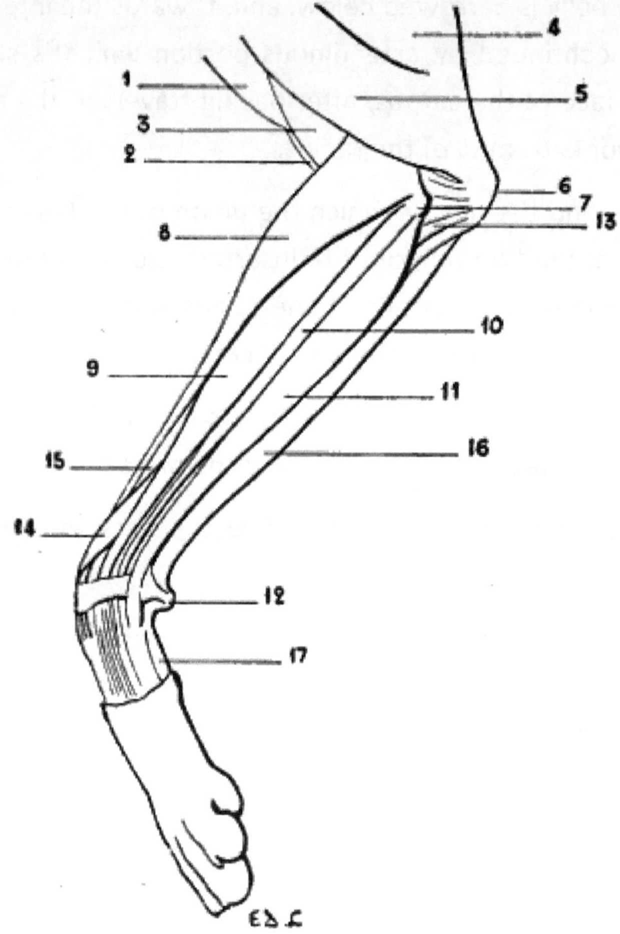

Fig. 73.—Myology of the Dog: Left Anterior Limb, External Aspect.

1, Mastoido-humeralis; 2, biceps; 3, brachialis anticus; 4, triceps, long portion; 5, triceps, external head; 6, olecranon process; 7, epicondyle; 8, radialis muscles (anterior extensor of the metacarpus); 9, extensor communis digitorum (anterior extensor of the phalanges); 10, extensor minimi digiti (lateral extensor of the phalanges, or common extensor of the three external digits); 11, posterior ulnar (external flexor of the metacarpus); 12, pisiform bone; 13, anconeus; 14, extensor ossis metacarpi pollicis and extensor primi internodii pollicis (oblique extensor of the metacarpus); 15, radius; 16, anterior ulnar (oblique flexor of the metacarpus); 17, external border of the hypothenar eminence (abductor of the little finger).

In order to properly understand and remember the respective positions occupied by these inferior insertions, it must be remembered that the human forearm being in the position of pronation, the tendons of the radials are attached to the bases of the metacarpals nearest to the thumb—that is to say, those occupying an internal position as regards the fourth and fifth metacarpals.

As its name indicates, this muscle extends the metacarpus. Consequently it is, in the horse, an extensor of the canon-bone.

It is also an adductor of the hand in those animals (cat, dog) in which the radio-carpal articulation, analogous in form to the corresponding articulation in man, permits lateral movements of the hand on the forearm. The union of the fleshy bodies of the two radials is sometimes found in the human species.

Supinator Brevis.—As in the case of the long supinator, the short supinator is found only in animals in which the radius can be rotated to a greater or less extent around the ulna; therefore this muscle is not found in the pig, the ox, or the horse; but it forms part of the forearm of the cat and the dog.

Deeply situated at the region of the elbow, the short supinator has little interest for us. All that we will say of it is that it goes from the external part of the inferior extremity of the humerus to the superior part of the radius; and that it is, in carnivora, the essential agent in the production of the movement of supination.

Extensor Communis Digitorum (Fig. 73, 9, 10, 11; Fig. 74, 10, 11, 12).—Also named in veterinary anatomy the *anterior extensor of the phalanges*, this muscle is situated external to and behind the anterior extensor of the metacarpus already described.

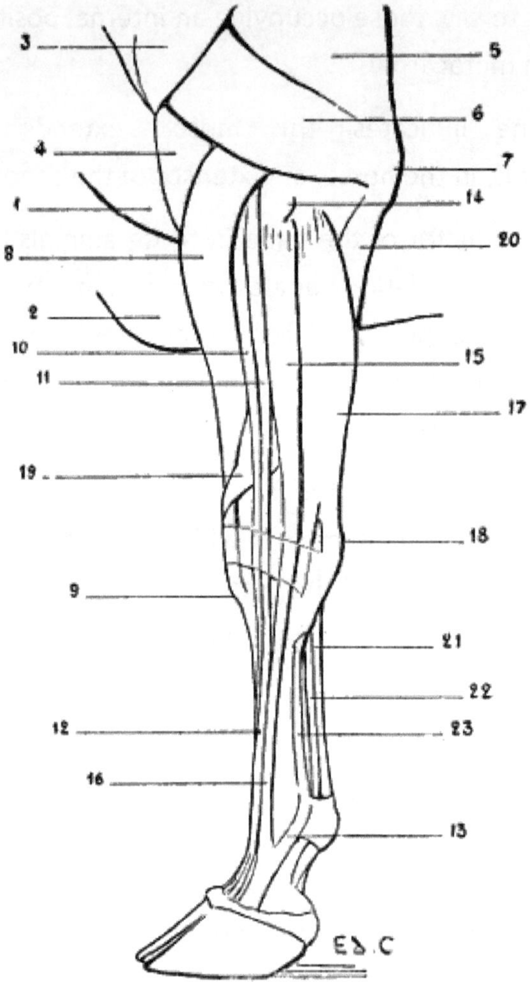

Fig. 74.—Myology of the Ox: Left Anterior Limb, External Aspect.

1, Mastoido-humeralis; 2, pectoralis major; 3, deltoid; 4, brachialis anticus; 5, triceps; 6, triceps, external head; 7, olecranon; 8, radial extensors (anterior extensor of the metacarpus); 9, insertion of the tendon of the anterior extensor of the metacarpus to the tubercle of the superior extremity of the principal metacarpal; 10, 11, extensor communis digitorum (10, proper extensor of the inner digits; 11, common extensor of the two digits); 12, tendon of the common extensor of the two digits; 13, band of reinforcement from the suspensory ligament of the fetlock; 14, external tuberosity of the superior extremity of the radius; 15, extensor minimi digiti (proper extensor of the external digit); 16, tendon of the proper extensor of the external digit; 17, posterior ulnar (external flexor of the metacarpus); 18, pisiform; 19, extensor ossis metacarpi pollicis and extensor primi internodii pollicis (oblique extensor of the metacarpus); 20, ulnar portion of the deep flexor of the toes; 21, tendon of the superficial flexor of the toes (superficial flexor of the phalanges); 22, tendon of the deep flexor of the toes (deep flexor of the phalanges); 23, suspensory ligament of the fetlock.

In the human being, the common extensor of the fingers springs, in its superior part, from the bottom of a depression, situated on the outer side of and behind the elbow, and limited in front by the muscular prominence which the long supinator and the first radial extensor form at that level. At the bottom of this hollow or fossette is found the epicondyle, which gives origin, amongst other muscles, to the common extensor of the fingers. It is necessary to add that it is most prominently visible during supination, and that it tends to be effaced during pronation.

An analogous arrangement is met with in animals. But the muscular prominence is formed by the united radial extensors, and the fossette, because of the permanent pronation of the forearm, is scarcely recognisable. Likewise, with regard to the dog, we may say that it does not exist, on account of the prominence which the epicondyle forms in that animal (Fig. 73, 7).

In connection with this prominence of the epicondyle, it is interesting to add that this detail recalls the relief which the same process produces on the external aspect of the human elbow when the forearm is flexed on the arm. We know that, in this case, the epicondyle is exposed, because the muscles which mask it in supination (long supinator and long radial extensor) are displaced and set it free during flexion. But, in the dog, as in other quadrupeds besides, the forearm is, in the normal state, flexed on the arm; the latter being oblique downwards and backwards, and the former being vertical. Further, the epicondyle is well developed.

The muscle with which we are now occupied, long and vertical in direction, arises from the inferior part of the external border of the humerus (there it is covered by the anterior extensor of the metacarpus, from which it is freed a little lower down) and from the external and superior tuberosity of the radius. In the carnivora, it arises from the epicondyle. Its fleshy body is fusiform in shape, becomes tendinous in the lower half of the forearm, and then divides into a number of slips, varying in number according to the species; this division is correlated to that of the hand—that is to say, with the number of the digits. Before reaching this latter, the common extensor of the digits passes through the most external groove on the anterior surface of the inferior extremity of the radius.

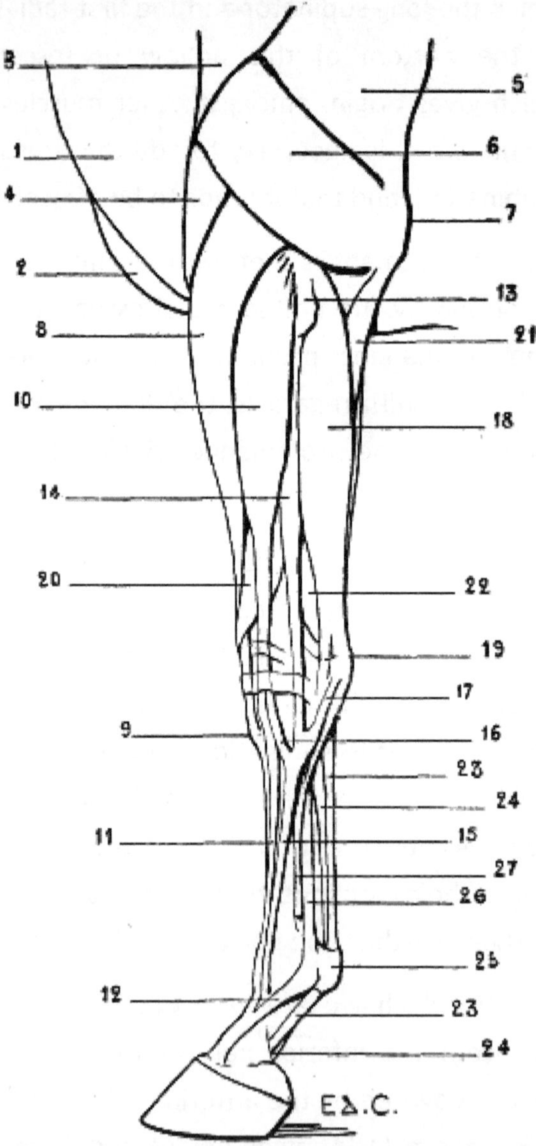

Fig. 75.—Myology of the Horse: Left Anterior Limb, External Aspect.

1, Mastoido-humeral; 2, pectoralis major; 3, deltoid; 4, brachialis anticus; 5, triceps, long head; 6, triceps, external head; 7, olecranon; 8, radial extensors (anterior extensor of the metacarpus); 9, insertion of the tendon of the anterior extensor of the metacarpus into the tubercle of the superior extremity of the principal metacarpal; 10, extensor communis digitorum (anterior extensor of the phalanges); 11, tendon of the anterior extensor of the phalanges; 12, reinforcing band arising from the suspensory ligament of the fetlock; 13, external tuberosity of the superior extremity of the radius; 14, extensor minimi digiti (lateral extensor of the phalanges); 15, tendon of the lateral extensor of the phalanges; 16, fibrous band which this latter receives from the tendon of the anterior extensor of the phalanges; 17, fibrous band which the same tendon receives from the carpal region; 18, posterior ulnar (external flexor of the metacarpus); 19, pisiform; 20, extensor ossis metacarpi pollicis and extensor primi internodii pollicis (oblique flexor of the metacarpus); 21, ulnar portion of the deep flexor of the digits; 22, flexor digitorum profundus; 23, 23, tendon of the flexor digitorum sublimis (superficial flexor of the phalanges); 24, 24, tendon of the flexor digitorum profundus (deep flexor of the phalanges); 25, sesamoid prominence; 26, suspensory ligament of the fetlock; 27, external rudimentary metacarpal.

In the cat and the dog, the four tendons which result from the division of the principal tendon go to the four last digits, and each of them is inserted, as in the human species, to the second and third phalanges.

In the pig, the anterior extensor of the phalanges is rather complicated in its arrangement. Its fleshy body is divided into four bundles terminated by tendons, which in turn divide and join certain digits; whence the special names given to each of these fasciculi, commencing with the most internal, of: *proper extensor of the great inner toe*; *common extensor of the two inner toes*; *common extensor of the two outer toes*; and *proper extensor of the great outer toe*.

In the ox, the same muscle is divided into two bundles: the internal proceeds to the internal toe, the external is common to the two toes.

In the horse, the tendon of the anterior extensor of the phalanges is divided into two parts of unequal bulk. The smaller of these tendinous slips, which is the more external, unites at the level of the superior part of the metacarpus with the tendon of the muscle which we are about to study in the following paragraph (Fig. 75, 16). The larger, after having reached the anterior surface of the digit, is attached to the anterior aspect of the first and second phalanges, and then forms a terminal expansion which is inserted into the pyramidal eminence of the third.

At the level of the first phalanx this tendon receives on each of its lateral aspects a strengthening band, which proceeds from the terminal extremity of *the suspensory ligament of the fetlock*,[26] and crosses obliquely downwards and forwards over the surface of the first phalanx to join the extensor tendon (Fig. 75, 12).

[26] See for a description of this ligament.

A similar arrangement is found in the ox.

This band is noticeable under the skin which covers the lateral aspects of the ham.

As the name indicates, this muscle extends the phalanges, one upon the other. It also contributes to the extension of the hand, as a whole, on the forearm.

Extensor Minimi Digiti (Fig. 73, 10; Fig. 74, 15, 16; Fig. 75, 14, 15).—This muscle, *the lateral extensor of the phalanges* of veterinary anatomy, situated on the external surface of the forearm, behind the common extensor of the digits, arises, as a rule, from the epicondyle (dog, cat), or from the external surface of the superior extremity of the radius (horse). The tendon succeeding to the fleshy body appears towards the lower third of the forearm, and at the level of the wrist lies in a groove analogous to that which in man is hollowed out for the passage of the corresponding tendon at the level of the inferior radio-ulnar articulation. This groove corresponds to the same articulation in animals in which the ulna is well developed, such as the dog and the cat; but it belongs to the radius when the inferior extremity of the ulna does not exist—for example, in the horse. Indeed, in this animal the groove in question is found on the external surface of the carpal extremity of the radius.

In the dog, the tendon is divided into three parts, which, crossing obliquely the tendons of the common extensor of the digits, pass to the three external digits, to be inserted by blending with the corresponding tendons of the latter into the third phalanges of those digits.

Thus is explained the name of *common extensor of the three external digits* which is sometimes given to this muscle.

In the cat, there is a fourth tendon, which passes to the index-finger, so that the name *common extensor of the four external digits* is in this case legitimate, and the lateral extensor of the phalanges is also a common extensor, as is the anterior extensor of the phalanges, or common extensor of the digits.

In the pig, the tendon, which is single, is inserted into the external digit, for which reason it has received the name of the *proper extensor of the small external digit*. This muscle is, then, really the homologue of that which exists in the human species.

In the ox, it is called the *proper extensor of the external digit*; it is as thick as the common extensor.

Finally, in the horse, the muscle is little developed. Its fleshy body, thin and flattened from before backwards, becomes distinctly visible only below the middle of the forearm. Above, it is enclosed in a limited space, bounded in front by the common extensor of the digits, and behind by the posterior ulnar; there these two muscles approach each other so closely that from the point of view of external form they seem to be nearly in contact.

The tendon, after receiving the small fasciculus from the common extensor (Fig. 75, 16), as well as a fibrous band emanating from the external surface of the carpus (Fig. 75, 17), is situated at the external side of the tendon of the anterior extensor of the phalanges, and is inserted into the anterior surface of the superior extremity of the first phalanx.

This muscle extends the digit or digits into which it is inserted. It also assists in the movement of extension of the hand as a whole.

Posterior Ulnar (*Extensor carpi ulnaris*) (Fig. 73, 11; Fig. 74, 17; Fig. 75, 18).—Designated by veterinary anatomists as the *external flexor of the metacarpus*,[27] or *external cubital*, this muscle is situated in the posterior region of the external surface of the forearm, behind the lateral extensor of the phalanges.

[27] Certain authors give it the name of *ulnar extensor of the wrist*. It is true that in the human being this is its action; but in quadrupeds, owing to its insertion into the pisiform, it draws the hand into the position of flexion.

It arises from the epicondyle; its fleshy body, thick but flattened, is directed vertically towards the carpus, and its tendon is inserted into the external part of the superior extremity of the metacarpus, after having given off a fibrous band, which takes its attachment on the pisiform.

It is inserted, in the cat and the dog, into the superior extremity of the fifth metacarpal; in the pig to the external metacarpal; in the ox to the external side of the canon-bone; in the horse to the superior extremity of the external rudimentary metacarpal.

This muscle flexes the hand on the forearm, and in animals in which the radio-carpal articulation permits, by its formation, it inclines the hand slightly outwards—that is, abducts it.

Anconeus (Fig. 72, 17; Fig. 73, 13).—We have already stated that the anconeus is included with the triceps brachialis in zoological anatomy, and that veterinary anatomists give it the name of *small extensor of the forearm*.[28]

[28] It is also called by some authors, the *small anconeus*.

In the dog it recalls, as to position, the human anconeus, but with this difference—that, in the latter, the anconeus, triangular in outline, has one of its angles turned outwards (the epicondyloid attachment) and one of its sides turned towards the olecranon. Here it is entirely the opposite. The anconeus, similarly triangular, is broader externally. At this level it takes its origin from the external border of the humerus, the epicondyle, and the external lateral ligament of the articulation of the elbow; thence its fibres converge towards the external surface of the olecranon, to be there inserted.

It is in relation, anteriorly and inferiorly, with the posterior ulnar muscle. It is covered superiorly by the external head of the triceps. In the cat the disposition of the anconeus is analogous. But in the other quadrupeds with which we are here concerned it is completely covered by the external head of the triceps. It really participates in the formation of the triceps; and seeing that it takes origin from the posterior surface of the humerus at the margin of the olecranon fossa (Fig. 72), and proceeds thence towards the olecranon to be inserted, we can understand why veterinary anatomists have connected its study with that of the posterior muscular mass of the arm.

This muscle is an extensor of the forearm on the arm.

We proceed now to inquire what the deep muscles of the posterior region of the human forearm become in quadrupeds: the long abductor of the thumb, the short extensor of the thumb, the long extensor of the thumb, the proper extensor of the index. We know that in every instance these muscles, which are deeply seated at their origin, become superficial afterwards.

In quadrupeds, on account of the position in which the forearm is placed—viz., pronation—the corresponding muscles occupy the anterior aspect of this region.

Long Abductor of the Thumb (*Extensor ossis metacarpi pollicis*) **and Short Extensor of the Thumb** (*Extensor primi internodii pollicis*) (Fig. 73, 14; Fig. 74, 19; Fig. 75, 20).—United one to the other in man, blended in quadrupeds, they form in the latter the muscles to which veterinary anatomists give the name of *oblique extensor of the metacarpus*.

This muscle arises from the median portion of the skeleton of the forearm. There it is covered by the common extensor of the digits and that of the small digit (anterior extensor and lateral extensor of the phalanges). Then, at the internal border of the first of these muscles, it becomes superficial, passes downwards and inwards, crosses superficially the anterior extensor of the metacarpus, reaches the inferior extremity of the radius, and becomes lodged in the most internal of the grooves situated on the anterior surface of this extremity, passes on the internal side of the carpus, and is inserted into the superior extremity of the most internal metacarpal—that is, to the first metacarpal, or metacarpal of the thumb—in the dog and cat; to the internal rudimentary metacarpal in the horse.

It is an extensor of the metacarpal into which it is inserted; but as, if we recall the extreme examples given above, in the dog the first metacarpal is not very mobile, and in the horse the internal rudimentary metacarpal is absolutely fixed to the bone which it accompanies, it is more exact to add that this muscle is principally an extensor of the metacarpus as a whole.

And yet, in the cat and the dog, it is also able to adduct the first metacarpal bone. It must be understood that this movement would be abduction, if the hand could be placed in the position of complete supination, as in the human species.

Long Extensor of the Thumb (*Extensor secundi internodii pollicis*) **and Proper Extensor of the Index** (*Extensor indicis*).—These two muscles are blended together by their fleshy bodies, so that the single name of *proper extensor of the thumb and index* is preferable. This muscle is but of slight importance from our point of view, for it is extremely atrophied, and so much the more as the number of the digits is lessened.

It arises, as the preceding, from the skeleton of the forearm, and there it is deeply placed. Below, towards the carpus, its tendinous part becomes superficial, to end in the following manner:

In the carnivora, the tendon divides into two very slender parts, which are inserted into the thumb and the index. In the pig, the tendon is blended with that of the common extensor of the internal digits. Finally, in the ox and the horse, it is sometimes regarded as being blended with the common or anterior extensor of the phalanges. But to us it appears more rational to say that it does not exist, which, moreover, is explained by the digital simplification of the hand.

Internal and Posterior Region

Pronator Teres (Fig. 76, 8).—This muscle, as may easily be understood, undergoes, as do the supinators, a degree of degeneration in proportion to the loss of mobility of the radius on the ulna. In animals in which the bones of the forearm are not fused it exists; in those, on the other hand, in which this segment has become simply a supporting column, it is not developed—at least, in a normal manner.

It is, consequently, found best marked in the dog and the cat.

Forming, as in man, the internal limit of the hollow of the elbow, the pronator teres has a disposition analogous to that which characterizes the corresponding muscle in the human species. It arises from the epitrochlea (internal condyle), proceeds downwards and outwards, and is inserted into the middle portion of the body of the radius.

It is into the hollow in front of the elbow, which this muscle contributes to limit, that the biceps and the brachialis anticus dip.

In the pig and the ox it is atrophied.

In the horse it does not exist. We may, however, sometimes find it, but in an abnormal form. We were able to demonstrate its presence in the form of a fleshy tongue situated on the internal side of the elbow (Fig. 78) in a horse which we dissected many years ago in the laboratory of the School of Fine Arts. Moreover—and the fact seemed to us an interesting one—the forearm to which the muscle belonged had an ulna of relatively considerable development (Figs. 79 and 80).[29]

[29] Édouard Cuyer, 'Abnormal Length of the Ulna and Presence of a Pronator Teres Muscle in a Horse' (*Bulletin de la Société d'Anthropologie*, Paris, 1887).

This muscle is a pronator.

Flexor Carpi Radialis (Fig. 76, 10; Fig. 77, 7).—Called by veterinary anatomists *the internal flexor of the metacarpus*, this muscle, which is found on the internal aspect of the forearm, is situated behind the pronator teres when this muscle exists, whilst in the animals which are deprived of the latter the flexor carpi radialis has in front of it the internal border of the radius, which separates it from the anterior extensor of the metacarpus.

It is necessary to add that the flexor carpi radialis is similarly separated from the anterior extensor of the metacarpus by the internal border of the radius in animals in which the pronator teres exists, but then only in that part of the forearm which is situated below this latter.

The flexor carpi radialis arises from the epitrochlea. Its fleshy body, fusiform in shape, descends vertically, and terminates in a tendon on the posterior surface of the bases of the second and third metacarpals in the dog and the cat, on the metacarpal of the large internal digit in the pig, on the internal side of the metacarpus in the ox, and on the superior extremity of the internal rudimentary metacarpal in the horse.

We see clearly, in this latter, a superficial vein which, in the shape of a strong cord, passes along the anterior border of the flexor carpi radialis; it is the subcutaneous median or internal vein, which, forming the continuation of the internal metacarpal vein, joins the venous system of the arm, after having crossed obliquely the corresponding part of the radius.

Palmaris Longus.—This muscle, which exists distinctly in some animals, but whose absence is far from being rare in the human species, is not developed as a distinct muscle in any of the domestic quadrupeds.

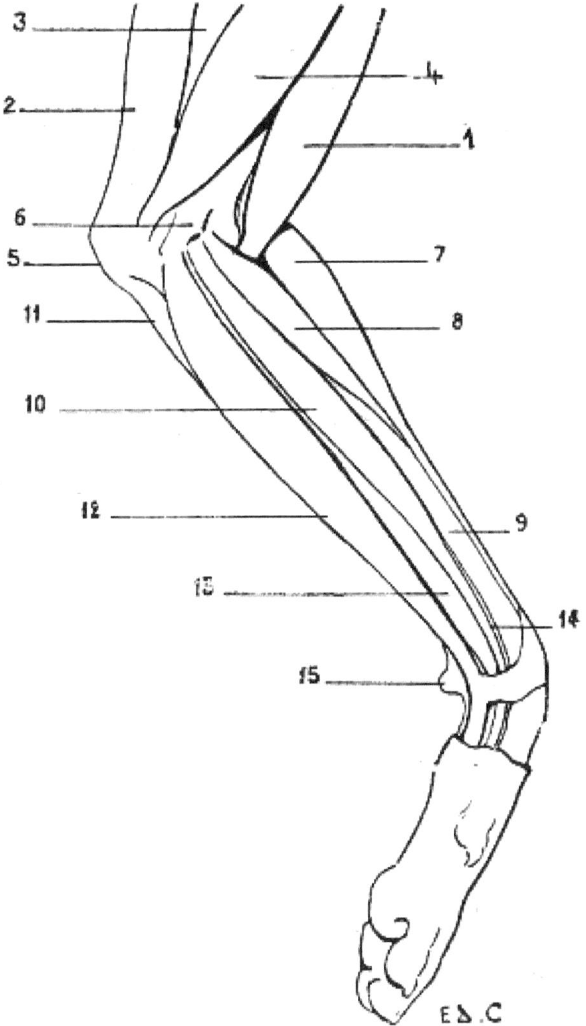

Fig. 76.—Myology of the Dog: Left Anterior Limb, Internal Aspect.

1, Biceps; 2, long extensor of the forearm (supplementary or accessory muscle of the great dorsal); 3, triceps, long head; 4, triceps, internal head; 5, olecranon; 6, epitrochlea (internal condyle); 7, radial extensors (anterior extensor of the metacarpus); 8, pronator teres; 9, radius; 10, flexor carpi radialis (internal flexor of the metacarpus); 11, anterior ulnar (oblique flexor of the metacarpus); 12, superficial flexor of the digits; 13, deep flexor of the digits; 14, flexor longus pollicis (radial fasciculus of the deep flexor of the digits); 15, pisiform bone.

And yet some authors announce its presence in the dog, and describe it as becoming detached, in the form of a cylindrical bundle, from the anterior surface of the fleshy mass of the deep flexor of the digits to proceed then by a tendon which divides into two parts, to terminate in the palm of the hand, where it blends with the tendons of the superficial flexor, which are destined for the third and fourth digits.

These authors give to this muscle the name of *palmaris longus*, and attribute to it the action of flexing the hand.

Anterior Ulnar (*Flexor carpi ulnaris*) (Fig. 73, 16; Fig. 76, 11; Fig. 77, 8).—Called by veterinary anatomists the *oblique flexor of the metacarpus*, or *internal ulnar*, this muscle occupies the internal part of the posterior aspect of the forearm in the ox and the horse, while in the dog it occupies rather the external part.

This difference arises from the fact that in this latter, as in man, the anterior ulnar is separated from the flexor carpi radialis by an interval in which we see, on the internal aspect of the forearm, just at the level of the elbow, the flexors of the digits. This interval is so much the wider as there is no palmaris muscle to subdivide its extent (Fig. 81). In the horse, the interval in question does not exist. In this animal, indeed, the anterior ulnar is in contact with the radial flexor, so that this muscle can occupy only a region belonging rather to the internal surface of the forearm (Fig. 82).

In the dog the anterior ulnar is in contact with the posterior ulnar. This relation recalls that which is found in man, where the two muscles are merely separated by the crest of the ulna (Fig. 81). But in the horse, in which the anterior ulnar has, so to speak, slid towards the internal aspect, this muscle is separated above from the posterior ulnar, and it is in the interval separating these two muscles that we are able to perceive, but this time at the back of the forearm, the muscular mass of the flexors of the digits (Fig. 82).

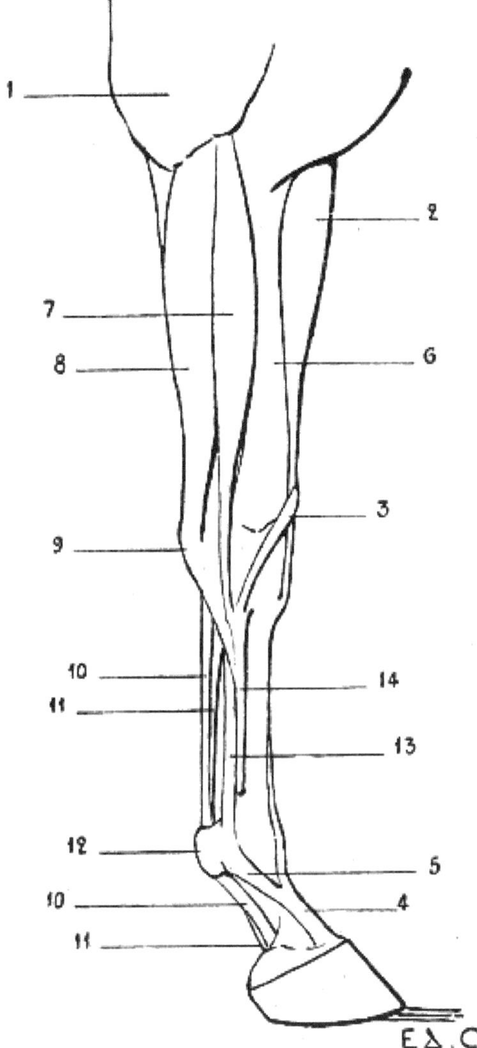

Fig. 77.—Myology of the Horse: Anterior Limb, Left Side, Internal Aspect.

1, Long extensor of the forearm (supplementary or accessory muscle of the latissimus dorsi); 2, radialis muscles (anterior extensor of the metacarpus); 3, tendons of extensor ossis metacarpi pollicis extensor primi internodii pollicis united (oblique extensor of the metacarpus); 4, tendon of extensor communis digitorum (anterior extensor of the phalanges); 5, strengthening band from the suspensory ligament of the fetlock; 6, internal surface of the radius; 7, flexor carpi radialis (internal flexor of the metacarpus); 8, anterior ulnar (oblique flexor of the metacarpus); 9, pisiform bone; 10, 10, tendon of the superficial flexor of the digits (superficial flexor of the phalanges); 11, 11, tendon of the deep flexor of the digits (deep flexor of the phalanges); 12, sesamoid prominence; 13, suspensory ligament of the fetlock; 14, internal rudimentary metacarpal.

The anterior ulnar arises above from the epitrochlea and the olecranon; thence it is directed towards the carpus, to be inserted into the pisiform bone. It proceeds therefore from the inner side of the elbow to the outer side of the upper part of the hand; it consequently crosses the posterior surface of the forearm obliquely. This is why, as we have pointed out above, it receives the name of the oblique flexor of the metacarpus.

It is not unprofitable to recall in this connection that there is an internal flexor of the metacarpus, which is the flexor carpi radialis; and an external flexor of the metacarpus, which is the posterior ulnar (in human anatomy, extensor carpi ulnaris). It is between these two muscles that we find the oblique flexor—the anterior ulnar which we have just been studying.

This muscle flexes the hand on the forearm.

Superficial Flexor of the Digits (*Flexor digitorum sublimis*) (Fig. 76, 12; Fig. 77, 10, 10).—This muscle arises from the epitrochlea; thence it passes towards the hand, becomes tendinous, passes in a groove on the posterior aspect of the carpus, and terminates on the palmar surface of the phalanges in furnishing a number of tendons proportioned to the digital division of the hand. Whatever the number, to which we will again refer, each tendon is attached to the second phalanx, after bifurcating at the level of the first, so as to form a sort of ring, destined to give passage to the corresponding tendon of the deep flexor. This ring and this passage have gained for the muscle the name of *perforated flexor*.

In the dog and the cat the principal tendon is divided into four parts, which go to the four last digits.

In the ox it is divided into two parts only; as, moreover, in the pig, whose superficial flexor is destined for the two large digits only, the lateral digits receiving no part of it.

Finally, in the horse the tendon is single.

We have previously pointed out that in the carnivora this muscle is visible on the internal and posterior aspects of the forearm, in the interval which is limited in front by the flexor carpi radialis and behind and outside by the anterior ulnar.

Certain details are still to be added to the description of this muscle. We will enter on an analysis of them after we have given some indications relative to the following muscle:

Deep Flexor of the Digits (*Flexor digitorum profundus*) (Fig. 75, 21, 22; Fig. 76, 12; Fig. 77, 11, 11).—This muscle is covered by the superficial flexor. It arises from the epitrochlea, from the radius, and from the ulna, either from the olecranon process—as in the ox, pig, and horse—or from almost the whole extent of the shaft of the same bone, as in the cat and dog.

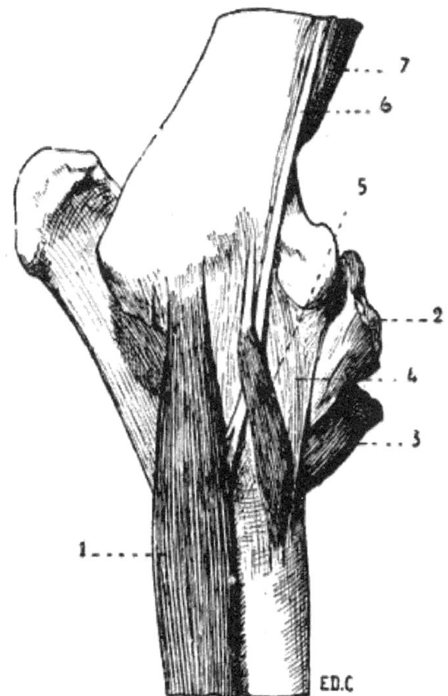

Fig. 78.—Left Anterior Limb of the Horse: Internal Aspect.

1, Internal flexor of the metacarpus or great palmar; 2, inferior part of the biceps; 3, inferior part of the brachialis anticus; 4, internal lateral ligament of the elbow; 5, pronato teres muscle.

The radial fasciculus represents in the domestic quadrupeds the long proper flexor muscle of the thumb in man. For this reason we shall describe the muscle afresh in the following paragraph:

The fleshy bundles of which we have just spoken terminate in a tendon which afterwards divides into slips, the number of which is in proportion to the digital division of the hand. These slips then pass through the slit or *buttonhole* in the tendon of the superficial flexor, and proceed to terminate on the third phalanx; hence the name of *perforating*, which is also given to the deep flexor of the digits.

In the dog and the cat the tendon is divided into five portions, each of which proceeds to one of the digits. The internal tendon, which is destined for the thumb, terminates on the second phalanx of this digit.

In the pig the tendon divides into four tendons destined for the four digits.

In the ox there are but two tendons.

In the horse the tendon is single.

As their names indicate, these muscles, both superficial and deep, flex the digits. In addition to this, they flex the hand on the forearm.

We mentioned above that certain details relative to the superficial flexor must be analyzed in a special way. We now add that this should also be done with regard to the deep flexor. The point in question is the arrangement which the tendons of these muscles present at the level of the palmar region of the hand.

It is easy, in the case of the dog or the cat, to picture to one's self this arrangement, especially if we recollect that which exists in the human species. The tendons of the flexors are placed on a kind of muscular bed formed by the union of the muscles of the region, but, moreover, from the point of view of external form, these tendons are not of very great importance.

But in the ox and the horse it is quite otherwise. From the simplification of the skeleton of the hand, and the reduction of the number of movements which the bones that form it are able to execute, there naturally results a diminution of its muscular apparatus. Apart from the existence of muscular vestiges of but little importance, we can say that, in reality, the hand does not possess any muscles. On its palmar aspect are found only the tendons of the flexors of the digits, and as these tendons are large, and the hand long, they give origin to external forms which it is necessary to examine.

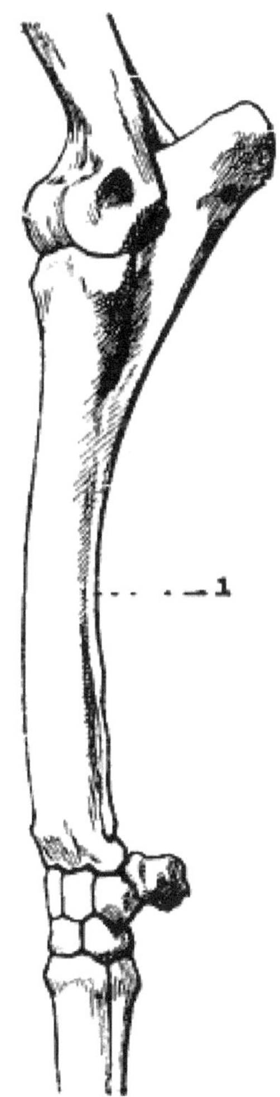

Fig. 79.—Left Anterior Limb of the Horse: External Aspect.

1, Ulna of abnormal length.

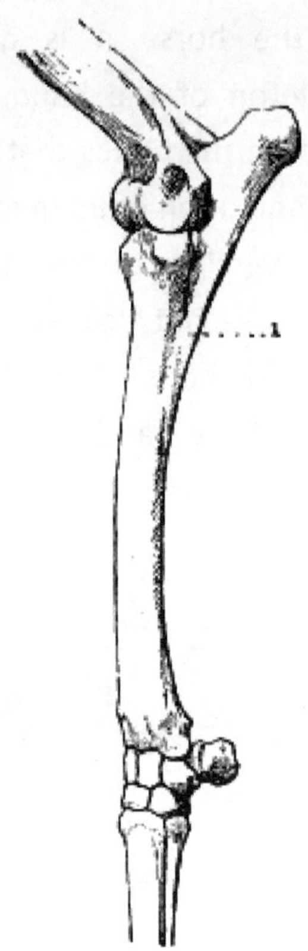

Fig. 80.—Left Anterior Limb of the Horse: External Aspect.

1, Normal ulna.

In the horse, which we take as a type, the tendons of the flexors, after being retained in position at the carpus by a fibrous band, the *carpal sheath*, which recalls the anterior annular ligament of the human carpus, and having passed this region, descend vertically, remaining separated from the posterior surface of the metacarpus, so that the skin sinks slightly on the lateral parts in front of the thick cord which these tendons form. This cord is known by the name of *tendon*.

The flexors then reach the fetlock, and occupy the groove formed by the peculiar arrangement of the two large sesamoid bones. They are retained in position at this level by a fibrous structure, which forms the metacarpo-phalangeal sheath. They then reach the phalanges, being directed obliquely downwards and forwards, as, moreover, the latter are also inclined. Then the tendon of the superficial flexor divides into two slips, which are inserted into the second phalanx, between which slips passes the tendon of the deep flexor, which in its turn goes to be inserted, in the form of an expansion, into the semilunar crest, by which the inferior surface of the third phalanx is divided into two parts.[30]

[30] See, as regards this crest, in the paragraph relative to the hoof of the solipeds, the figures which represent the third phalanx, viewed on its inferior surface (Figs. 101 and 102).

The part which these tendons play is of great importance in the large quadrupeds.

These tendons, in fact, in addition to the action determined by the contraction of the fleshy fibres to which they succeed, maintain the angle formed by the canon-bone and the phalangeal portion of the hand, and prevent its effacement under the weight of the body during the time of standing. Their strong development, and the position they occupy, make this understood, without it being necessary to insist on it further.

We mentioned above that the 'tendon' descends vertically from the carpus towards the fetlocks. This is as it should be. But, in some horses, it is oblique downwards and backwards, so that the canon, instead of being of equal depth from before backwards in its whole length, is a little narrower in its upper part.

This results from the fact that the tendons of the flexors, too firmly bound by the carpal sheath, gradually separate as they pass from the metacarpus, going to join the fetlock; hence the obliquity pointed out above. This abnormality producing a deleterious result, in the sense that the tendinous apparatus acts with less strength as an organ of support, it constitutes a defect of conformation which is expressed by saying that the tendon has 'failed.'

Long Proper Flexor of the Thumb (*Flexor longus pollicis*) (Fig. 76, 14).—As we have already pointed out, this muscle is represented in quadrupeds by the radial bundle of the deep flexor of the digits, so that the two muscles are in reality blended the one to the other. This union is sometimes found, but only as an abnormality, in the human species. We have met some examples of this in the course of our dissections.

Pronator Quadratus.—This muscle conforms to the general law which we have already pointed out in connection with those which have for their action the rotation of the radius around the ulna. We remember, indeed, that when the bones of the forearm are fused with one another, the muscles which are destined to produce a mobility which has then become impossible disappear at the same blow.

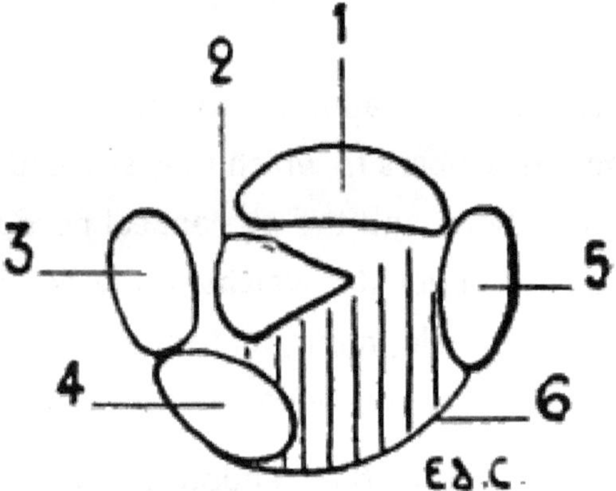

Fig. 81.—Diagram of the Posterior Part of a Transverse Section passing through the Middle of the Left Fore-limb of the Dog: Surface of the Inferior Segment of the Section.

1, Radius; 2, ulna; 3, posterior ulnar; 4, anterior ulnar; 5, great palmar (*flexor carpi radialis*); 6, flexors of the digits.

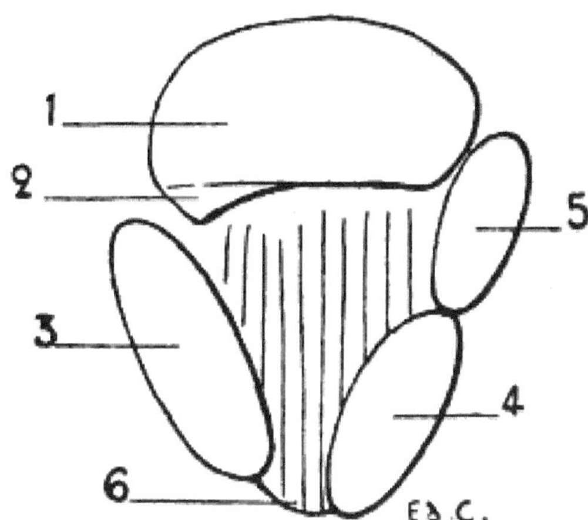

Fig. 82.—Diagram of a Horizontal Section of the Middle of the Forearm of the Left Leg of the Horse: Surface of the Inferior Segment of the Section.

1, Radius; 2, ulna; 3, posterior ulnar; 4, anterior ulnar; 5, great palmar (*flexor carpi radialis*); 6, flexors of the digits.

For this cause we do not find the square pronator in either the ox or the horse, but can demonstrate its presence in the dog and the cat.

It is very deeply situated. This is why, and also on account of the plan which we have traced for ourselves, we will simply say that it is situated on the postero-internal aspect of the skeleton of the forearm, and that it extends from the ulna to the radius.

It seems to us, however, sufficiently interesting to add that, instead of occupying, as in the human species, the inferior fourth of the two bones, it extends, particularly in the dog, over their whole length, with the exception of their superior and inferior extremities.

Muscles of the Hand

We will first recall that, in man, the palm of the hand is divided into three regions: a median palmar region, which is occupied by the tendons of the flexors of the digits, the lumbricales, and, deeply, by the interosseous muscles; an external region, or thenar eminence, formed by the muscles destined for the movements of the thumb; an internal region, or hypothenar eminence, which contains the muscles proper to the small digit and the palmar cutaneous muscle.

These muscles are found, more or less developed, in the dog and the cat.

In the ox and the horse we meet with no vestige of the muscles of the thenar or hypothenar eminences. Nevertheless, in these animals we find the muscles which belong to the central region of the palm. We refer to the lumbricales and the interosseous.

Although this fact has no relation to the object of our study, it appears to us interesting to announce that there are traces of the lumbricales found in the solipeds. These muscles are represented by two fleshy bundles, situated one on each side of the tendon of the deep flexor, above the ring of the tendon of the superficial flexor. These small muscles are continued as slender tendons, which become lost in the fibrous tissue of the *spur*, which is the horny process situated at the posterior part of the fetlock, and which is covered by the hairs, more or less abundant, which constitute the *wisp*.

As for the interosseous muscles, they are represented by the *suspensory ligament of the fetlock*, and by two other small muscles, tendinous throughout, which are situated between the principal metacarpal and the rudimentary ones.

The suspensory ligament of the fetlock is considered an interosseous muscle, on account of the red fleshy striations which it contains, and from certain relations which it forms with the tendon of the common extensor of the digits or anterior extensor of the phalanges. This ligament (Fig. 75, 26; Fig. 77, 13), which plays an important part in the standing position as a support of the foot, is a fibrous band situated between the tendons of the flexors of the digits and the principal metacarpal. It arises above, from the second row of the carpals, descends towards the fetlock, where it divides into two branches, which are inserted into the large sesamoid bones. At the same level, this ligament gives off two fibrous bands which, passing downwards and forwards, join the tendon of the anterior extensor of the phalanges, blending with it, after having each crossed one of the lateral aspects of the pastern. We have already referred to these bands.

It is with these latter that are blended the long and slender tendons which form in a great measure the two other interosseous muscles previously described.

A ligament of the same kind is found in the ox (Fig. 74, 23).

MUSCLES OF THE POSTERIOR LIMBS

Muscles of the Pelvis

The muscles which specially interest us in this region, because of their superficial position, are the gluteus maximus and the gluteus medius. As for the gluteus minimus, it is deeply situated, and more or less sharply marked off from the second of the preceding muscles.

Inasmuch as the gluteus medius is more simple in arrangement than the maximus, and will aid us in arranging our ideas in connection with the latter, it is with the study of it that we will commence.

Gluteus Medius (Fig. 68, 29; Fig. 69, 26; Fig. 70, 35).—This muscle, as in man, occupies the external iliac fossa. But this latter being directed differently in the digitigrades and the ungulates, as we have pointed out in the section on osteology, the muscle in question has consequently not the same direction in the two groups of animals, being turned outwards in the first, and upwards in the second.

It is the thickest of the glutei, and gives to the region which it occupies a rounded form.

From the iliac fossa from which it arises the fleshy fibres are directed towards the femur, to be inserted into the great trochanter. It is covered by an aponeurosis, and in part by the great gluteal. It completely covers the small gluteal, which veterinary anatomists designate by the name of the *deep gluteal*.

In the carnivora it does not pass in front of the iliac crest, but, in the ox, and more particularly in the horse, it is prolonged anteriorly, and thus covers, to a certain extent, the muscles of the common mass.

When it contracts, taking its fixed point at the pelvis, the gluteus medius extends the thigh, which it is also able to abduct. If, on the other hand, its fixed point is on the femur, it acts on the trunk, which it raises, producing oscillating movements of the pelvis. It contributes in this way to the action of rearing. We also see it distinctly appear by the prominence which it produces in the dog, which, according to the time-honoured phrase, *fait le beau*.

Gluteus Maximus (Fig. 68, 28; Fig. 69, 25; Fig. 70, 33, 34).—The great gluteal muscle, further designated in veterinary anatomy the *superficial gluteal*, is proportionately less developed in quadrupeds than in man. Indeed, in the latter, where it is of very great thickness, its volume is due to the important function which it fulfils in maintaining the biped attitude.

In quadrupeds it contributes to form the superficial part of the crupper and the external surface of the thigh. It is divided into two parts: one anterior, the other posterior.

With regard to this latter, it will be necessary to indicate how it has been sometimes regarded, and to what portion of the muscular system in man it corresponds. But we believe that it is better to see beforehand, without any preconceived idea, how these two parts are arranged.

In the dog, the anterior portion of the gluteus maximus arises from the sacrum, while some fibres situated further forward arise from the surface of the gluteus medius, near the iliac spine, and from the tensor of the fascia lata with which these fibres are blended. The posterior portion, united to the preceding—that is to say, to those of its fibres which arise from the sacrum—takes its origin from the first coccygeal vertebra. These two portions are directed towards the femur, to be inserted into the great trochanter, and to the external branch of the superior bifurcation of the linea aspera.

In the cat, the posterior bundle is less definitely blended with the anterior. By a long and slender tendon which, behind, turns around the great trochanter, and passes along the surface of the fascia lata, it proceeds to join the knee-cap.

In the pig, the posterior portion is much more developed.

In the horse, the anterior portion arises from the internal iliac spine (posterior in man), from the external iliac spine (anterior in man), and, between these two osseous points, from the aponeurosis, which covers the gluteus medius. Between these two origins the muscle is deeply grooved, so that the tendency is to divide into two portions, each of which is directed towards one of the iliac angles. In this groove the gluteus medius is to be seen.

The fleshy bundles converge, and are directed towards the external aspect of the femur, to be inserted into the osseous prominence known as the third trochanter, after passing beneath the fleshy fibres of the posterior portion. The latter, which is more considerable than the preceding portion, arises above from the sacral crest, from the aponeurosis which envelops the coccygeal muscles, from the sacro-sciatic ligament, and from the tuberosity of the ischium. From this origin it passes downwards, expands, then, describing a curve with the convexity behind, it becomes narrowed, and proceeds to be inserted by a deep fasciculus into the third trochanter, to the fascia lata, and, lastly, to the knee-cap by the inferior part of its tendon.

Above, its posterior border is covered by the semi-tendinosus; interiorly, the same border is in relation with the biceps femoris.

In the ox, the two parts of the great gluteal muscle are blended together.

The long and broad fleshy band which they form arises in a manner corresponding to that which we have just indicated in connection with the horse, except that it has no attachment to the femur. The fascia lata adheres strongly to its anterior border for a considerable length. The form of the superior border of the great gluteal muscle of this animal differs from that of the analogous portion in the horse. This difference results from the peculiar aspect which the corresponding region of the pelvis presents, and from the fact that, in the ox, as the semi-tendinosus does not cover the portion of the great gluteal which arises from the tuberosity of the ischium, the attachments of this muscle to the sacro-sciatic ligament are uncovered.

Its descending portion, as a whole, has a rectilinear form, and does not form a curve such as we indicated in the case of the horse.

The anterior portion of the great gluteal flexes the thigh. As regards the posterior portion, it extends the thigh, and abducts it.

The action of this latter portion is particularly interesting as regards the horse, because of the great development of the muscular mass which this region presents in this animal. If the muscle takes its fixed point above, it acts, in the extension of the thigh during walking, by projecting the trunk forward during the whole time that the hind-limb to which it belongs is in contact with the ground. If, on the contrary, it takes its fixed point below, it makes the pelvis describe a see-saw movement, upwards and backwards, on the coxo-femoral articulation, and so contributes to the action of rearing.

Now that we have a knowledge of the disposition of the great gluteal muscle, the moment has come to inquire what is the signification of its posterior portion. The action of the anterior part being clearly comparable to the human great gluteal, there can be no doubt as regards the homology of this portion, so we will not insist on it further.

Of the posterior portion it is wholly different, for it is the homologue of a fleshy bundle annexed to the great gluteal of man, but which is not developed except as an abnormality.

Indeed, we sometimes find, placed along the inferior border of the great gluteal, a fleshy fasciculus, separated from this muscle by a slight interspace. This fasciculus, long and narrow, takes origin from the summit of the sacrum, or the coccyx, and goes to partake of the femoral insertions of the muscle which it accompanies. We further note a muscle of the same kind, and presenting the same aspect, which comes from the tuberosity of the ischium. Notwithstanding the difference which exists, it is this abnormal fasciculus of man which in the quadrupeds here studied is considered as constituting the posterior portion of the great gluteal.

Bourgelat, considering this posterior portion as belonging to the biceps cruris, to which, it is true, it adheres, forms of them a muscle which he designates under the name of the *long vastus*. The anterior fasciculus of this long vastus is none other than the posterior portion of the great gluteal which we have just been studying.

Muscles of the Thigh

These muscles are divided into three regions: posterior, anterior, and internal.

In a corresponding manner to that which we described in connection with the arm, the thigh is applied to the side of the trunk, and is free, more or less, only at the level of the inferior part.

Further, by reason of this shortening of the femur, the great gluteal muscle, which is elongated in the ox and the horse, for example, occupies in part the region corresponding to that which in man is occupied by the muscles of the thigh, which here are reduced in length. In other words, they are not superposed, as in the human species, but juxtaposed. This is what we will verify further on.

The thigh, as a whole, is flattened from without inwards, its transverse diameter being less in extent than its antero-posterior. Its external surface is slightly rounded; that is, of course, in quadrupeds with sufficiently well-developed muscles. Its internal surface is known as the *flat of the thigh*.

Muscles of the Posterior Region

It is not unprofitable to recall to mind what muscles form the superficial layer of this region in the human being. They are the biceps cruris, semi-tendinosus, and semi-membranosus. We now proceed to discover their analogues in quadrupeds.

Biceps Cruris (Fig. 68, 30; Fig. 69, 27; Fig. 70, 36).—It is this which, according to Bourgelat, forms the central and posterior portions of the long vastus muscle which we have mentioned above.

We know that the biceps of man is so named from the two portions which form its upper part. In domestic quadrupeds, and also in the majority of the mammals, this muscle is reduced to a single portion, that which comes from the pelvis. It is therefore the portion which arises from the femur which does not exist. This condition is sometimes found as an abnormality in the human species.

The biceps arises from the tuberosity of the ischium; hence it is directed, widening as it goes, towards the leg, where it terminates by an aponeurosis which blends with the fascia lata and the aponeurosis of the leg, and then proceeds to be attached to the anterior border or crest of the tibia. By its inferior portion it limits externally the posterior region of the knee—the popliteal space.

A fibrous intersection traverses the biceps in its whole length, with the result that the muscle looks as if formed of two portions, one of which is situated in front of the other.

In the dog and the cat it also arises from the sacro-sciatic ligament. At this level its contour is distinguishable from that which corresponds to the gluteal muscles, so that we there find two prominences one above the other. The superior is formed by the gluteal muscles; the inferior corresponds to the tuberosity of the ischium. The two prominences are separated by a depression, from which the biceps emerges. We draw attention to this form, the character of which is so expressive of energy in the carnivora.

In these animals the biceps is inserted, by its anterior fibres, into the articulation of the knee, while in the rest of its extent it covers in great measure by its aponeurosis the external aspect of the leg.

In the pig, the biceps is but slightly marked off from the posterior part of the great gluteal. In the ox, the division between these two muscles is a little more distinct.

In the horse, the sciatic origin of the biceps is covered by the semi-tendinosus, so that it only becomes free lower down, to appear in the space limited behind by the semi-tendinosus, and in front by the posterior part of the gluteus maximus.

When the biceps contracts, taking its fixed point from above, it flexes the leg and helps to extend the thigh. If, on the other hand, it takes its fixed point from below, it lowers the ischium, makes the pelvis undergo a see-saw movement, and acts thus in the movement of rearing. It is sometimes called, on account of one of its actions, and the position which it occupies, the 'external flexor, or peroneal muscle of the leg.'

Semi-tendinosus (Fig. 68, 31; Fig. 70, 37; Fig. 87, 1; Fig. 88, 1; Fig. 89, 28).—This muscle forms the contour of the thigh posteriorly, so that when the latter is viewed from the side, it is the semi-tendinosus above all that forms the outline. But, as we shall soon see, it is in this case more distinct above than below, because of the deviation which it undergoes in order to occupy by its inferior part the internal side of the leg.

In the dog, the cat, and the ox, the semi-tendinosus arises from the tuberosity of the ischium only, as in the human species. In the pig, it also takes origin higher up from the sacro-sciatic ligament and the coccygeal aponeurosis. In the horse, it extends still further, for it is also attached to the crest of the sacrum.

The indication of these origins is of importance from the point of view of external form, and to convince ourselves of this it is sufficient to compare, in the ox and the horse, the region of the pelvis situated below the root of the tail. In the ox, whose semi-tendinosus arises from the tuberosity of the ischium only, this region is depressed, and the cavity which is formed at this level is limited behind by the tuberosity, which we know is very thick and prominent above. This causes the superior part of the crupper to be less oblique than in the horse. This characteristic is more especially marked in the cow, the bull having this region of a more rounded form.

In the horse, on account of the semi-tendinosus ascending to the coccyx, and even to the sacrum, the depression in question does not exist, and the presence of the tuberosity of the ischium is only slightly revealed.

Descending from the origin indicated above, and inclining more and more inwards, the semi-tendinosus proceeds to blend with the aponeurosis of the leg, to be inserted into the anterior border of the tibia, after crossing over the internal surface of the latter. It forms the internal boundary of the popliteal space.

When this muscle contracts, taking its fixed point at the pelvis, it flexes the leg. If, on the other hand, it takes its fixed point at the tibia, it makes the pelvis describe a see-saw movement, and acts accordingly in the movement of rearing.

It is sometimes named the 'internal or tibial flexor of the leg,' in opposition to the crural biceps, which, as stated above, is then the external flexor of the same region.

Semi-membranosus (Fig. 68, 32; Fig. 87, 2; Fig. 88, 2).—This muscle, situated on the inner side of the semi-tendinosus, can be seen only when the thigh is regarded on its posterior aspect.

It is only by reason of the homology of situation with the corresponding muscle in man that we give the name under which we are studying it; indeed, its structure is different, for it does not present the long, broad, aponeurotic tendon which, in its superior part, characterizes this muscle in the human species.

It arises above from the inferior surface of the ischium, and from the tuberosity of the same bone. In the pig, and especially in the horse, it passes further upwards, to arise from the aponeurosis of the coccygeal muscles. So that if we compare it with that of the ox, which does not extend beyond the ischium, we find that it is associated with the semi-tendinosus in determining the difference of aspect to which we have already called attention in connection with the region of the pelvis situated below the root of the tail.

The semi-membranosus is then directed downwards and forwards, to take its place on the internal surface of the thigh, where it is partly covered by the gracilis muscle. It is inserted in the following manner:

In the dog and the cat it is divided into two parts, anterior and posterior. The first, the more developed, is attached to the internal surface of the inferior extremity of the femur; the second to the internal tuberosity of the tibia.

The same arrangement occurs in the ox.

In the horse it is inserted into the internal surface of the internal condyle of the femur.

The semi-membranosus is an extensor of the thigh when it takes its fixed point at the pelvis; it is also an adductor of the lower limb. If it takes its fixed point below it assists in the action of rearing.

It is now necessary for us, especially as regards the horse, to add some indications relative to the exterior forms of the region constituted by the semi-membranosus and semi-tendinosus. These two muscles form, by their union, a surface contour, slightly projecting and of elongated form, which occupies the posterior border of the thigh, the contour corresponding to the region known as the *buttock*, in spite of the fact that none of the gluteal muscles take any part in the structure of this region. But the appearances, to a certain extent, justify the preservation of this name. Indeed, because of the groove which separates the gluteal region of one side from that of the opposite side, and from the position of the anal orifice in the superior part of this groove, we may admit the name which, in hippology, has been given to this part of the thigh.

In addition to the reasons just given, and which are justified especially by the position occupied by the muscular mass formed by the union of the two muscles, there is another which, this time, has a relation to a certain detail of form. In the superior part of the convexity, which the gluteal region describes in the greater part of its extent, there is found a more salient point, greatly accentuated in lean animals, due to the presence of the tuberosity of the ischium; it is the *point* or *angle of the buttock*. At this level, and near the median line, the semi-membranosus, not aponeurotic, but fleshy, and even thicker there than anywhere else, sometimes produces a sharply localized prominence. And as this prominence is situated on the outer side of the anal orifice, the resemblance to a small 'buttock' is still more marked.

In lean horses a deep groove separates the mass formed by the semi-membranosus and semi-tendinosus from that of the other muscles of the thigh situated more in front; this groove is known by a name which in this case is remarkably expressive—that of the 'line of poverty.'

If we examine the gluteal region as a whole by looking at the thigh from the side, we plainly see the graceful curve produced by the general convexity above indicated. We return to this point, in order to add that, in its lower part, this curve alters its character; that is to say, it is replaced by a slight concavity. This, which is designated under the name of *the fold of the buttock*, is situated close to the level of articulation of the leg with the thigh-bone.

Muscles of the Anterior Region

First we recall that in man the anterior muscles of the thigh are: the triceps cruris, the tensor of the fascia lata, and the sartorius.

Triceps Cruris (Fig. 8, 36; Fig. 69, 31; Fig. 70, 41; Fig. 84, 2; Fig. 87, 3; Fig. 88, 3).—This muscle, which occupies the greater part of the space between the pelvis and the anterior aspect of the femur, consists of three parts: an external, or vastus externus; an internal, or vastus internus; and a median or long portion, or rectus femoris. This division accordingly recalls that which characterizes the human triceps cruris. Furthermore, as in the case of the latter, the vastus externus and the vastus internus take their origin from the shaft of the femur, while the long portion arises from the pelvis. The *vastus externus* arises from the external lip of the linea aspera of the femur (or from the external border of the posterior surface of this bone in the ox and the horse, in which the linea aspera, considerably widened, especially in the latter, forms a surface), and from the external surface of the shaft of the femur. From this origin its fibres pass downwards and forwards, to be inserted into the tendon of the long portion of the muscle and into the patella.

In the dog and the cat the vastus externus is the most voluminous of the three portions which constitute the triceps muscle. It is covered by the fascia lata; but notwithstanding this, its presence is revealed by a prominence which occupies the external surface of the thigh, and surmounts, in the region of the knee, the more slightly developed one which is produced by the knee-cap.

The *vastus internus*, situated on the inner surface of the thigh, takes its origin from the corresponding surface of the femur, and proceeds towards the patella.

The rectus femoris arises from the iliac bone, above the cotyloid cavity; its fleshy body, which is fusiform, and situated in front of and between the two vasti muscles, is directed towards the patella, into which it is inserted by a tendon, which receives the other two portions.

It is covered in front by the tensor of the fascia lata, and contributes with the vastus externus to form the upper prominence of the knee.

The ligamentous fibres, which, as in man, unite the knee-cap to the tibia, transmit to this latter the action determined by the contraction of the triceps. This muscle is an extensor of the leg. Furthermore, the rectus femoris, or long portion, acts as a flexor of the thigh.

Tensor Fascia Lata (Fig. 68, 34, 36; Fig. 69, 30, 31; Fig. 70, 40).—This muscle, generally larger in quadrupeds than in man, is flat and triangular, and occupies the superior and anterior part of the thigh.

It arises from the anterior iliac spine (inferior in carnivora, external in the ox and the horse); it is prolonged downwards by an aponeurosis (fascia lata) which occupies the external aspect of the thigh, proceeds to be inserted into the patella and blend with the aponeurosis of the biceps muscle.

It covers the rectus and vastus externus portions of the triceps cruris; it is also in relation with the gluteal muscles.

The tensor of the fascia lata flexes the thigh, and serves to raise the lower limb as a whole.

Sartorius (Fig. 68, 35; Fig. 87, 4, 5; Fig. 88, 5).—This muscle, long and flattened, is called by veterinarians *the long adductor of the leg*.

Before beginning the study of its position in quadrupeds, it is necessary to remember that in man, where the thigh has a form almost conical, the sartorius commences on the anterior face of this latter, and is directed downwards and inwards to reach the internal surface of the knee.

But now let us suppose the thigh flattened from without inwards; there will evidently result from this a change in situation with regard to the muscle in question. In fact, when this supposition is admitted, it is easy to imagine that in a great part of the extent in which the sartorius is normally anterior it will become internal. This is why, these conditions being realized in quadrupeds, we shall find that, in some of them, the sartorius is situated on the aspect of the thigh which is turned to the side of the trunk.

In the dog and the cat it arises from the anterior iliac spine, and from the half of the border of the bone situated immediately below it; but the fibres from this second origin being hidden by the tensor of the fascia lata, on the inner side of which they are situated, viewing the external surface of the thigh, the muscle seems to arise from the iliac spine only.

The sartorius in these animals is divided into two parts, which, in general, are placed in contact. One of these fasciculi is anterior; the other is situated further back. The first is visible on the anterior border of the thigh, in front of the tensor of the fascia lata, but below it inclines inwards; in its superior part also, a small extent of the internal surface is occupied by it. The second, which, as we have said, is situated further back, belongs wholly to the inner surface of the thigh; it is this portion which arises from the inferior border of the ilium (this is the homologue of the anterior border of the human iliac bone).

The two fasciculi then pass towards the knee, being in relation with the rectus and the vastus internus of the triceps. The anterior fasciculus is inserted into the patella. The posterior unites with the tendons of the gracilis (see below) and semi-tendinosus, and then proceeds to be inserted into the superior part of the internal surface of the tibia.

On account of their different insertions these two parts receive the names of *the patellar sartorius* and *tibial sartorius* respectively.

In the ox and the horse the sartorius is still more definitely situated on the internal surface of the thigh. Consisting of a single fasciculus, representing the tibial sartorius of the cat and the dog, it arises in the abdominal cavity from the fascia covering the iliac muscle, then passes under the crural arch, and terminates, by an aponeurosis which blends with that of the gracilis, on the inner fibres of the patellar ligament. In short, the sartorius is of interest to us in the carnivora only, and especially on account of its anterior or patellar fasciculus.

It is an adductor of the leg and a flexor of the thigh.

Muscles of the Internal Region

The ilio-psoas pectineus and the adductors which we study in man, in connection with the internal aspect of the thigh, offer little of interest from the point of view of external form in quadrupeds; it is for this reason that we will disregard them.

The gracilis alone merits description.

Gracilis (Fig. 87, 9; Fig. 88, 6).—Designated in veterinary anatomy under the name of *the short adductor of the leg*, this muscle, expanded in width, occupies the greater part of the internal surface of the thigh, *or flat of the thigh*, as this region is also called. Let us imagine, in man, the internal surface of the thigh broader, and the internal rectus more expanded, and we shall have an idea of the same muscle as it exists in quadrupeds.

The gracilis arises from the ischio-pubic symphysis and from the neighbouring regions; thence it is directed towards the leg to be inserted into the superior part of the internal surface of the tibia, after being united to the tendons of the sartorius and semi-tendinosus. We find, accordingly, at this level, an arrangement which recalls the general appearance of what in man receives the name of *the goose's foot* (*pes anserinus*).

It is between this muscle and the sartorius, at the superior part of the internal surface of the thigh, in the region which recalls the triangle of Scarpa, that we are able, especially in the cat and the dog, to see the adductor muscles of the thigh. We also partly see there, in these animals, the vastus internus and the rectus of the triceps (see Fig. 87). The gracilis is an adductor of the thigh.

Muscles of the Leg

We will divide the leg into three regions: anterior, external, and posterior. With regard to the internal region, there are no muscles which belong exclusively to it; for it is in great measure formed by the internal surface of the tibia, which, as in man, is subcutaneous.

Muscles of the Anterior Region

We first note that in the human species the tibialis anticus, extensor proprius pollicis, extensor longus digitorum and the peroneous tertius or anticus, form the subcutaneous layer of this region. We now proceed to study these muscles in quadrupeds.

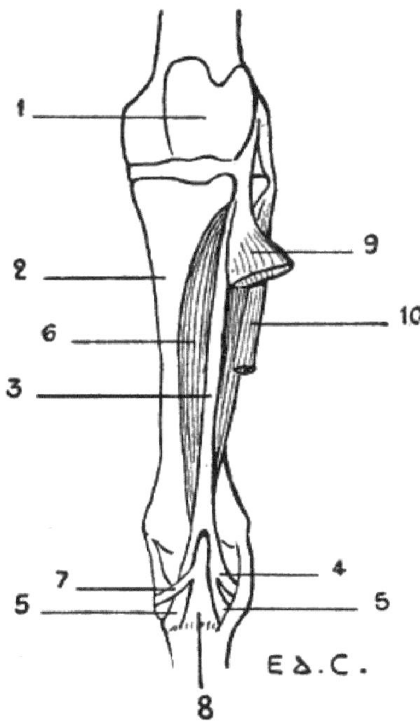

Fig. 83.—Myology of the Horse: the Anterior Tibial Muscle (Flexor of the Metatarsus), Left Leg, Anterior View.

1, Femoral trochlea; 2, tibia; 3, tendinous portion of the tibialis anticus; 4, cuboid branch of same; 5, 5, its metatarsal branch; 6, fleshy portion; 7, cuneiform branch of its tendon; 8, metatarsal branch of the same tendon; 9, extensor longus digitorum (anterior extensor of the phalanges turned outwards); 10, peroneus brevis (lateral extensor of the phalanges).

Tibialis Anticus (Fig. 83; Fig. 84, 6; Fig. 85, 4; Fig. 87, 10; Fig. 88, 10, 11).—It is further named by veterinarians the *flexor of the metatarsus*.

In the dog and the cat this muscle, which is rather large, arises from the external tuberosity of the tibia and from the crest of this bone. In its superior part it is flat, but lower down it is thick and produces a prominence in front of the tibia. Finally, it becomes tendinous, and passes towards the tarsus; thence it is directed towards the inner side of the metatarsus, and is inserted into the great-toe, this latter being sometimes well developed, but also often merely represented by a small bony nodule on which the muscle is then fixed.

In the other animals with which we here occupy ourselves, the tibialis anticus presents a complexity which would be incomprehensible unless this muscle be first studied in the horse.

In this latter the tibialis anticus consists of two distinct portions, placed one in front of the other: a fleshy portion, and a tendinous portion running parallel to it.

The muscle is covered, except on its internal part and inferiorly, by a muscle with which we will occupy ourselves later on—that is, the common extensor of the toes.

The tendinous portion of the tibialis anticus (Fig. 83), especially covered by the extensor of the toes, arises from the inferior extremity of the femur, from the fossa situated between the trochlea and the external condyle; thence it descends towards a groove which is hollowed out on the external tuberosity of the tibia, and is directed towards the tarsus, where it divides into two branches, which are inserted into the cuboid bone and the superior extremity of the principal metatarsal. These two parts form a ring through which the terminal tendon of the fleshy portion of the same muscle passes.

This fleshy portion, situated behind the preceding, arises from the superior extremity of the tibia, on the borders of the groove in which the tendinous portion lies; thence it passes downwards for a short distance on the inner side of the common extensor of the toes, which covers it in the rest of its extent. It ends in a tendon which, after passing through the tendinous ring above noticed, divides into two branches. One of these branches is inserted into the anterior surface of the superior extremity of the principal metatarsal, the other into the second cuneiform bone.

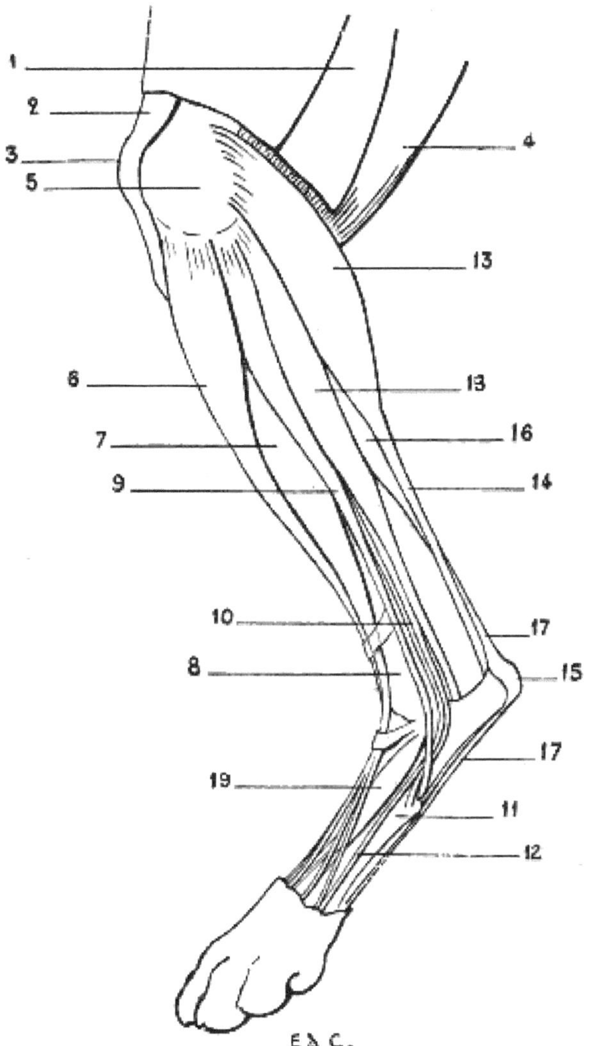

Fig. 84.—Myology of the Dog: Left Hind-limb, External Aspect.

1, Biceps cruris and fascia lata, divided in order to expose the upper part of the muscles of the leg; 2, inferior portion of the triceps cruris; 3, patella; 4, semi-tendinosus; 5, inferior extremity of the femur; 6, tibialis anticus (flexor of the metatarsus); 7, extensor longus digitorum (anterior extensor of the phalanges); 8, tibia; 9, peroneus longus; 10, peroneus brevis; 11, fifth metatarsal;

12, fasciculus detached from the peroneus brevis and passing towards the fifth toe; 13, external head of gastrocnemius; 14, tendo-Achillis; 15, calcaneum; 16, flexor digitorum sublimis; 17, 17, tendon of the flexor digitorum sublimis; 18, flexor longus pollicis (portion of the deep flexor of the toes); 19, dorsal muscle of the foot (short extensor of the toes).

In the ox the same two portions of the tibialis anticus exist, but with this capital difference—that the anterior portion is fleshy, superficial, and blended for a great part of its length with the common extensor of the toes.

The portion which corresponds to that which is fleshy in the horse arises from the tibia; below, it ends on the inner surface of the superior extremity of the metatarsus and the cuneiform bones. That which represents the tendinous part, which is also fleshy, as we have just pointed out, arises above with the common extensor of the toes, from the femur, in the fossa situated between the trochlea and the external condyle; whilst below, after having given passage to the tendon of the preceding portion, as in the horse, it is inserted into the metatarsus and the cuneiform bones.

In the pig, the tibialis anticus presents an arrangement nearly similar to that which we have just described.

It seems to us of interest to add that it has been sought to ascertain to what muscle of the human leg the tendinous part of the tibialis of the horse corresponds—a part which has become fleshy in the pig and the ox.

According to some authors, it represents the peroneus tertius; but that muscle is situated on the outer side of the common extensor of the toes; and here the portion with which it has been compared is placed on the inside. It has also been likened to a portion of the common extensor of the toes, but it does not pass to the latter. Lastly, it has been considered as being the homologue of the proper extensor of the great-toe; but why, then, in the ox, which has no great-toe, is it so highly developed? Nevertheless, its position and its relations sufficiently warrant this method of comprehending it. The tibialis anticus is a flexor of the foot. It is also able, in animals in which the tarsal articulations allow of the movement, to rotate the foot inwards.

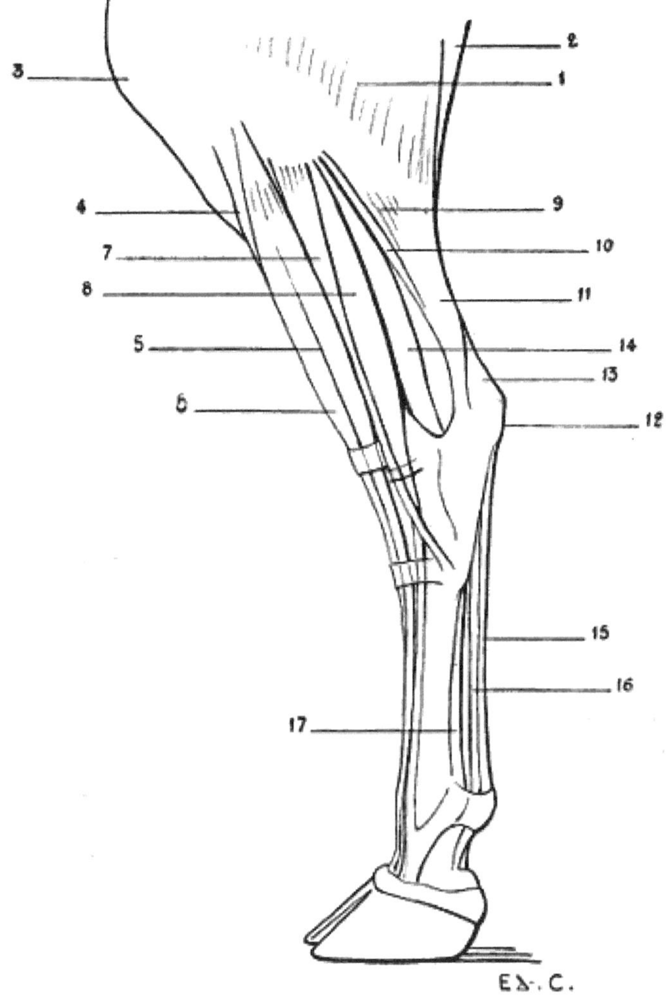

Fig. 85.—Myology of the Ox: Left Leg, External Aspect.

1, Gluteus maximus and biceps cruris; 2, semi-tendinosus; 3, patella; 4, tibialis anticus (flexor of the metatarsus); 5, extensor longus digitorum (anterior extensor of the phalanges); 6, fasciculus of the extensor longus digitorum, which is considered as the representative of the tendinous portion of the tibialis anticus in the horse; 7, peroneus longus; 8, peroneus brevis (proper extensor of the external toe); 9, external head of gastrocnemius; 10, soleus; 11, tendo-Achillis; 12, calcaneum; 13, tendon of the extensor longus digitorum (superficial flexor of the phalanges); 14, flexor longus pollicis and tibialis posticus (deep flexor of the phalanges); 15, tendon of the superficial flexor of the toes; 16, tendon of the deep flexor of the toes; 17, suspensory ligament of the fetlock.

With regard to the tendinous part, called by veterinarians the *cord of the flexor of the metatarsus*, it serves, in the horse, to produce the flexion of the metatarsus when the knee is already flexed; it thus acts in a passive fashion, which is explained by its resistance and the position which it occupies in relation to these two articulations.

Extensor Proprius Pollicis.—This muscle exists only in the dog and the cat, and there in a rudimentary condition.

It is covered by the common extensor of the toes and the tibialis anticus, and passes, accompanied by the tendon of this latter muscle, to terminate on the second metatarsal, or the phalanx, which articulates with it. When the first toe exists in the dog, it is inserted into this by a very slender tendon.

Extensor Longus Digitorum (Fig. 83, 9; Fig. 84, 7; Fig. 85, 5, 6; Fig. 86, 4; Fig. 87, 12; Fig. 88, 7).—It is also called by veterinarians *the anterior extensor of the phalanges*.

In the dog and the cat this muscle is to be seen in the space limited behind by the peroneus longus and in front by the tibialis anticus. Above it is covered by this latter. In the lower half of the leg, it is also in relation, on the inner side, with the tibialis anticus; but behind it is separated from the peroneus longus by the external surface of the shaft and inferior extremity of the tibia. This arrangement, besides, recalls that which is found in man, the peroneus longus of the latter diverging in the same way, at this level, from the common extensor, and leaving exposed the corresponding portion of the skeleton of the leg.

This muscle, fusiform in shape, arises at its upper part from the external surface of the inferior extremity of the femur, then its tendon passes into a groove hollowed on the external tuberosity of the tibia. The fleshy body which succeeds is directed towards the tarsus, but before reaching it is replaced by a tendon. This tendon, at the level of the metatarsal bones, divides into four slips, which pass towards the toes, and are inserted into the second and third phalanges of the latter.

In the horse it covers, to a great extent, the tibialis anticus, so that it is the latter which forms the large fusiform prominence especially noticeable in the middle region, to which the contour of the anterior surface of the leg is due.

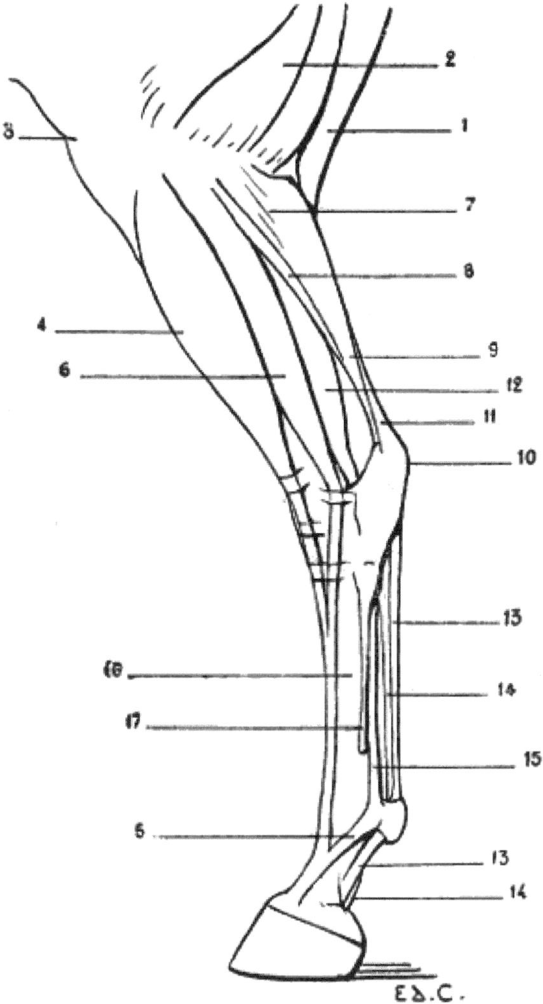

Fig. 86.—Myology of the Horse: Left Hind-limb, External Aspect.

1, Semi-tendinosus; 2, biceps cruris; 3, patella; 4, extensor longus digitorum (anterior extensor of the phalanges); 5, reinforcing band arising from the ligament of the fetlock; 6, peroneus brevis (lateral extensor of the phalanges); 7, external head of gastrocnemius; 8, soleus; 9, tendo-Achillis; 10, calcaneum; 11, tendon of the superficial flexor of the toes (superficial flexor of the phalanges); 12, flexor longus pollicis and tibialis posticus (deep flexor of the phalanges); 13, 13, tendon of the superficial flexor of the phalanges; 14, 14, tendon of the deep flexor of the phalanges; 15, suspensory ligament of the fetlock; 16, principal metatarsal: 17, external rudimentary metatarsal.

It arises above from the inferior extremity of the femur, from the fossa situated between the trochlea and the external condyle; therefore, it has a common origin with the tendinous portion of the tibialis anticus, or flexor of the metatarsus.

The tendon, which at the level of the inferior part of the leg succeeds to the fleshy body, passes in front of the tarsus, the principal metatarsal, and receives the tendon of the peroneus brevis which we will describe later on. It then reaches the anterior surface of the fetlock. There it presents an arrangement analogous to that which we have pointed out in connection with the anterior extensor of the phalanges—a muscle which, in the fore-limbs, corresponds to the common extensor of the digits; that is to say, it is inserted, in form of an expansion, into the pyramidal prominence of the third phalanx, after having formed attachments to the first and second, and having received on each side a strengthening band from the suspensory ligament of the fetlock.

In the ox the long extensor of the toes is united above, and for a great part of its length, with the portion of the tibialis anticus, which represents, albeit in the fleshy state, the tendinous cord of the latter in the horse.

In common with this portion, it arises from the inferior extremity of the femur. Thence it passes towards the tarsus and divides into two fasciculi, internal and external, which are continued by tendons. These pass towards the phalanges, and, in case of the common extensor of the digits belonging to the fore-limbs, the internal is destined for the internal toe, and the external is common to the two toes.

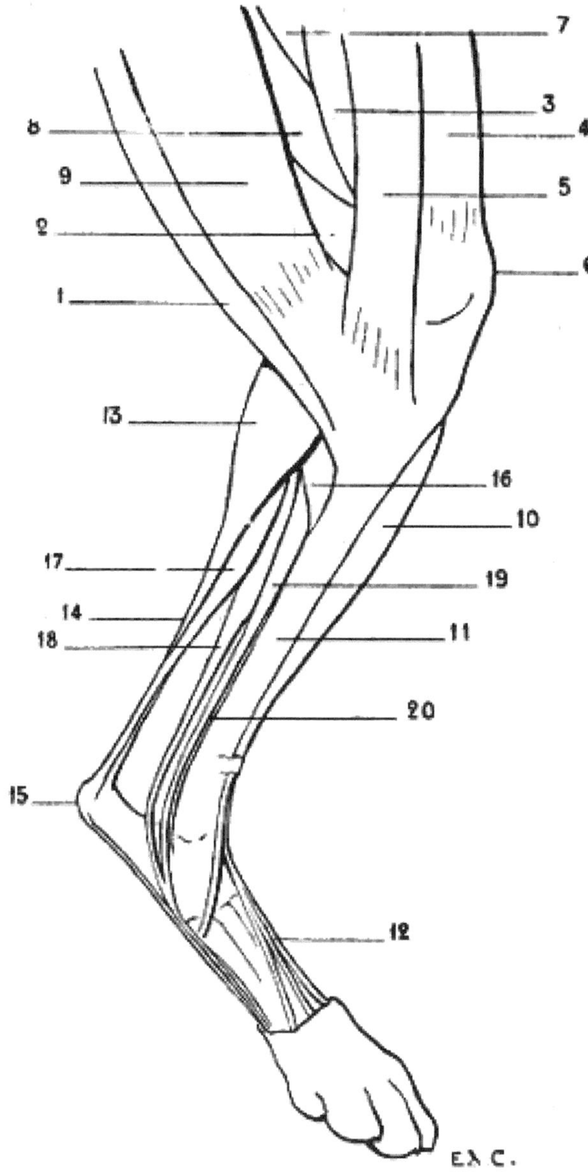

Fig. 87.—Myology of the Dog: Left Hind-limb, Internal Aspect.

1, Semi-tendinosus; 2, semi-membranosus; 3, triceps cruris (vastus internus); 4, sartorius (patellar); 5, sartorius (tibial); 6, patella; 7, first or middle adductor; 8, small and great adductor united; 9, gracilis; 10, tibialis anticus (flexor of the metatarsus); 11, tibia; 12, tendon of extensor longus digitorum (anterior extensor of the phalanges): 13, gastrocnemius, inner head; 14, tendo-Achillis; 15, calcaneum; 16, popliteus; 17, superficial flexor of the toes; 18, flexor longus pollicis (portion of the deep flexor of the toes); 19, flexor longus digitorum (portion of the deep flexor of the toes); 20, tendon of the tibialis posticus.

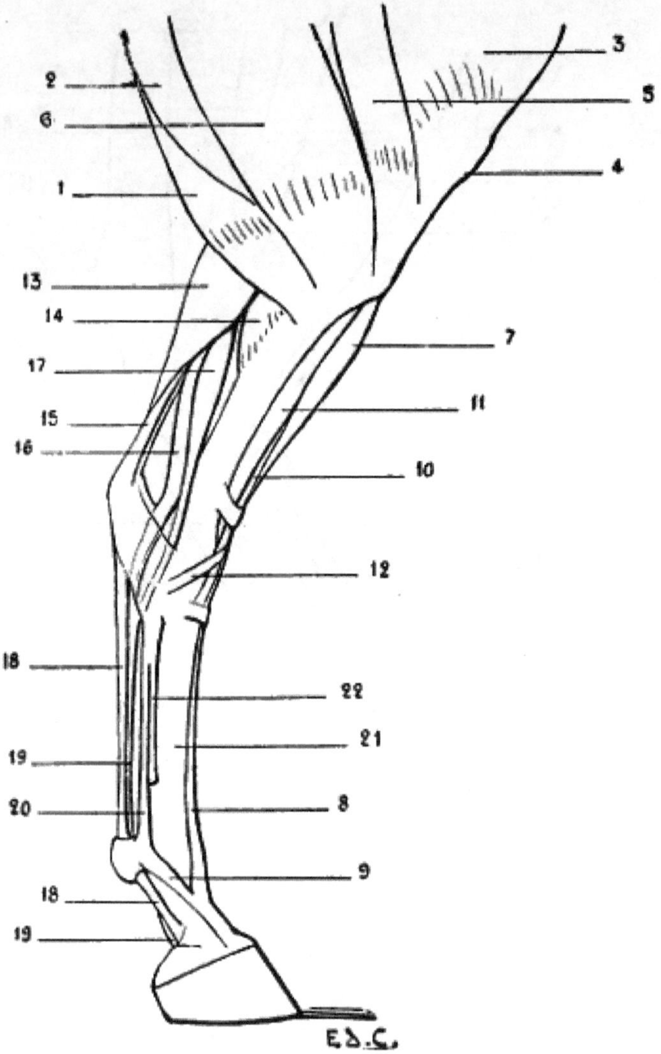

Fig. 88.—Myology of the Horse: Left Hind-leg, Internal Aspect.

1, Semi-tendinosus; 2, semi-membranosus; 3, triceps cruris (vastus internus); 4, patella; 5, sartorius; 6, gracilis; 7, extensor longus digitorum common extensor of the toes (anterior extensor of the phalanges); 8, tendon of the preceding muscle; 9, reinforcing band given off by the suspensory ligament of the fetlock; 10, tibialis anticus (flexor of the metatarsus), its tendinous portion; 11, tibialis anticus (flexor of the metatarsus), its fleshy portion; 12, cuneiform branch of the tendon of this fleshy portion; 13, internal head of gastrocnemius; 14, popliteus; 15, tendon of the flexor brevis digitorum (superficial flexor of the phalanges); 16, flexor longus pollicis and tibialis posticus (deep flexor of the phalanges); 17, flexor longus digitorum (oblique flexor of the phalanges); 18, 18, tendon of the superficial flexor of the phalanges; 19, 19, tendon of the deep flexor of the phalanges; 20, suspensory ligament of the fetlock; 21, principal metatarsal; 22, internal rudimentary metatarsal.

In the pig the general arrangement of the muscle is similar, but the tendons end in a manner which is a little more complicated. Apart from the fasciculi which correspond to the tendinous portion of the tibialis anticus (fleshy here, as in the ox), the long extensor of the toes at the level of the tarsus divides into three tendons: the internal goes to the great internal toe; the middle bifurcates in the upper part of the digital portion of the foot, and each of its branches goes towards one of the great-toes; the external divides to pass towards each of the two small toes, and towards the great ones; but this latter disposition is not constant.

By its contraction the muscle which we have just studied extends the phalanges and flexes the foot.

Peroneus Tertius.—This muscle is not found in domestic quadrupeds.

We should remember, nevertheless, that certain authors consider as representing it the tendinous portion of the anterior tibial of the horse, or the corresponding portion now fleshy, of the same muscle in the pig and the ox. It is by reason of this fact that it is called the third peroneal, notwithstanding that in the numerical order of the peroneals it is rather the first.

But that which still further complicates this question of nomenclature is that some authors give this name of third to a peroneal which, in the carnivora, is situated more definitely in the group of external muscles (see below, **Short Lateral Peroneal**).

Muscles of the External Region

In man, two muscles constitute this region; they are the peroneus longus and peroneus brevis.

Peroneus Longus (Fig. 84, 9; Fig. 85, 7).—This muscle does not exist in the domestic animals; only in the flesh-eaters, the pig and the ox excepted.

It is in relation superiorly with the tibialis anticus, and inferiorly with the common extensor of the toes; in the ox, it is in contact with this latter muscle throughout its whole length.

The peroneus longus arises from the external tuberosity of the tibia; towards the middle of the leg it is replaced by a tendon. This proceeds towards the tarsus, but previously it passes between the tibia and fibula. In the ox it is situated in front of the coronoid tarsal bone; we recollect that this bone is regarded as representing the inferior extremity of the fibula. Then it passes into a groove belonging to the cuboid bone or to the cuboido-scaphoid bone in the ox, traverses obliquely the posterior aspect of the tarsus, and is inserted into the rudimentary bone which represents the first toe; or, if this does not exist, into the innermost of the metatarsal bones.

This muscle is an extensor of the foot. It also rotates it outwards in the animals in which the articulation permits this latter movement.

Peroneus Brevis (Fig. 83, 8; Fig. 83, 10; Fig. 84, 10; Fig. 86, 6).—In the dog and the cat, this muscle is covered in part by the peroneus longus, and arises from the inferior half of the tibia and the fibula; at the level of the tarsus it becomes tendinous, passes into a groove hollowed out on the external surface of the inferior extremity of the fibula, and terminates on the external aspect of the superior extremity of the fifth metatarsal. A little before this insertion it crosses the tendon of the long peroneal in passing to the outer side of the latter.

To the short peroneal muscle is found annexed a very thin fasciculus which lies upon it. This fasciculus arises from beneath the head of the fibula, and is soon replaced by a thin tendon, which, accompanying that of the short peroneal, proceeds towards the foot, after having traversed the groove in the inferior extremity of the fibula; passes along by the fifth metatarsal (Fig. 84, 12); blends at the level of the first phalanx of the fifth toe with the corresponding tendon of the long extensor of the toes, and partakes of the insertions of this tendon.

This fasciculus is designated by some authors under the name of the peroneal of the fifth toe, or the proper extensor of the same toe. But what makes still further complications is that other authors regard it as an anterior, or third, peroneal. Now, these names are those which other anatomists have applied to the fasciculus of the anterior tibial, which, in the pig and the ox, is fused in part with the long extensor of the toes. Hence there results a confusion which is truly regrettable.

In brief, we can, without inconvenience, consider it as a fasciculus of the short peroneal muscle.

We sometimes find in man, but abnormally, an arrangement which partly recalls that which we have just indicated. It consists in a duplication of the tendon of the short peroneal, one of the branches of which goes to the fifth metatarsal, and the other to the fifth toe; it is sometimes a single fasciculus which goes to the phalanges of this latter. We have met with examples of these anomalies.[31] In the pig, the short peroneal is situated on the same plane as the long. It consists of two clearly distinct fasciculi, which arise from the fibula. The tendon of the anterior fasciculus proceeds to the great external toe—that is to say, the fourth, of which it is the proper extensor. The posterior fasciculus terminates on the small external toe, the fifth, of which it is in like manner the extensor.

[31] Édouard Cuyer, 'Anomalies, Osseous and Muscular' (*Bulletins de la Société d'Anthropologie*, Paris, 1891).

In the ox, the fleshy fibres of the short peroneal arise from a fibrous band which replaces the fibula, and from the external tuberosity of the tibia. Situated behind the long peroneal and on the same plane, it terminates in a tendon which appears at the level of the inferior part of the leg; it passes in front of the canon, and is inserted into the external toe, of which it is the proper extensor.

In the horse, it is the sole representative of the peroneal muscles, and veterinary anatomists have given it the name of *the lateral extensor of the phalanges*.

Its fleshy body arises above from the external lateral ligament of the knee-joint, and from the whole length of the fibula. In the middle third of the leg it is narrowed; lower down it is replaced by a tendon. This is lodged in a groove hollowed on the external surface of the inferior extremity of the tibia; then after passing along the external surface of the tarsus, it is directed forward, and proceeds to blend towards the middle of the canon-bone with the tendon of the long extensor of the toes, or anterior extensor of the phalanges, of which it shares the insertions. It extends the phalanges into which it is inserted. It also flexes the foot.

Muscles of the Posterior Region

It will not be unprofitable to recall to mind that, in man, the muscles of this region are arranged in two layers: a superficial layer consisting of the gastrocnemius and soleus, to which is added a muscle of little importance, the plantaris, and a deep layer formed by four muscles—the popliteus, flexor longus digitorum, tibialis posticus, and flexor longus pollicis.

The gastrocnemius and soleus, independent in their upper portion, unite below in a common tendon; they thus form also a triceps muscle, which we designate under the name of the triceps of the leg, or triceps suralis, because it forms the elevation of the calf of the leg (from *sura*, calf).

Gastrocnemius (Fig. 83, 9, 11; Fig. 84, 13, 14; Fig. 86, 7, 9; Fig. 88, 13).—The external and internal heads of the gastrocnemius, distinct from one another only in their upper portion, arise from the shaft of the femur, above the condyles, on the borders of the popliteal surface, to a relatively considerable extent in the great quadrupeds.

At this level they are situated in the popliteal region—that is to say, in the space limited externally by the biceps, and internally by the semi-tendinosus. But as they descend to a rather low level on the leg in quadrupeds, and especially in carnivora, they do not, properly speaking, determine a projection of the calf of the leg. However, they pass from this region but to be soon continued by a tendon—the tendo-Achillis, which is inserted into the calcaneum.

Now, the region of the tarsus is called by veterinarians *the ham*, the posterior surface of which is angular, because of the oblique direction of the leg with regard to the vertical direction of the metatarsus and the presence of the calcaneum; the prominence which this surface presents has received the name of *the point of the ham*, and the tendon which ends there that of *the cord of the ham*.

But the tendo-Achillis does not alone form this cord. Indeed, as we will soon see, the tendon of the superficial flexor of the toes takes part in its formation.

We may add, with regard to the tendo-Achillis, that it is more clearly perceived as an external feature, because the skin sinks in front of it, as it does in man, over the lateral parts of the region which it occupies.

The gastrocnemius, when it contracts, extends the foot on the leg.

It serves to maintain the tibio-tarsal angle in the standing position, and during walking, to determine the steadying of the hind-limbs, which then, after the fashion of a spring, project the body forward.

By an analogous movement they take part in the posterior projection of the hind-limbs in the act of kicking; but they are not the only ones to act in this case, the muscles of the buttock and thigh also being brought into play.

Soleus (Fig. 83, 10; Fig. 86, 8).—This muscle, much less developed in quadrupeds than in man, does not exist in the dog.

With regard to the soleus in the pig, Professor Lesbre says: 'Meckel denied its existence; we, however, believe that it is united to the external head of the gastrocnemius, its origin being transferred to the femur.'[32]

[32] F. X. Lesbre, 'Essai de Myologie comparée de l'homme et des mammifères domestiques en vue d'établir une nomenclature unique et rationelle,' Lyon, 1897, p. 169.

But in animals in which it exists, this muscle, of but little importance, occupies the outer side of the leg. It arises above from the external tuberosity of the tibia, and terminates below in a tendon which is united with that of the gastrocnemius.

The soleus has the same action as these latter.

Plantaris.—In quadrupeds this muscle is blended with the superficial flexor of the toes, which we will study afterwards.

Popliteus (Fig. 87, 16; Fig. 88, 14).—In man, this muscle, which occupies the posterior surface of the tibia, above the oblique line, is completely covered by the gastrocnemius.

In quadrupeds, where it is more voluminous, it projects internally beyond the gastrocnemius, so that it is seen in the internal and superior part of the region of the superficial layer of muscles, immediately behind the internal surface of the tibia, which, as we know, is subcutaneous.

The popliteus arises from the external surface of the external condyle of the femur. Thence its fibres which diverge pass to be inserted into the superior part of the posterior surface and of the internal border of the tibia. It is in this latter region that it projects beyond the gastrocnemius, but we may add that there it is more or less covered by the semi-tendinosus.

It flexes the leg, and rotates it forwards.

Superficial Flexor of the Toes (Fig. 83, 13, 15; Fig. 84, 17; Fig. 86, 11, 13, 13; Fig. 87, 17; Fig. 88, 15, 18, 18).—In man, the homologue of this muscle is found in the sole of the foot. It is called *the short flexor of the toes*. It arises from the calcaneum, and passes to the four outer toes. In quadrupeds, it rises as high as the back of the knee, and is found blended with the plantaris.

Further designated by the name of *the superficial flexor of the phalanges*, covered in part by the gastrocnemius, with which it is in relation for a great part of the course which it traverses, this muscle arises from the posterior surface of the femur, on the external branch of the inferior bifurcation of the linea aspera. In the horse, this origin takes place in a depression situated above the external condyle, in the supracondyloid fossa. Then it accompanies the gastrocnemius, and becomes tendinous where the tendo-Achillis commences. It then winds round the latter in placing itself on its inner side, then on its posterior surface, and reaches the calcaneum. It accordingly contributes, as we have already pointed out, to form the cord of the ham. After having become expanded, and having covered as with a sort of fibrous cap the bone of the heel, it descends behind the metatarsus, and presents there an arrangement analogous to that which we pointed out in connection with the superficial flexor of the digits—that is, it contributes to form the *tendon*. This prominence, in the form of a cord, we see behind the canon-bone in solipeds and ruminants. It finally terminates in the same way as the muscle with which we have compared it.

In the horse, its fleshy body is but slightly developed, so that its tendon alone is specially visible in the superficial muscular layer, but in the dog and the cat it is large. Hence it results that its fleshy body appears on each side of the inferior half of the gastrocnemius, and produces an elevation which recalls that which the soleus produces on each side of the same muscles in the human species.

The muscles which follow form, with the popliteus, which we have already studied, the deep layer of the posterior region of the leg.

Flexor Longus Digitorum (Fig. 87, 19; Fig. 88, 17).—This muscle, in man, is the only common flexor of the toes belonging to the muscles of the leg.

In comparison with the preceding muscle, it is a deep flexor. Veterinarians have given it the name of *the oblique flexor of the phalanges*.

Visible on the internal aspect of the superficial layer of the muscles of the leg, this muscle arises above from the posterior surface of the external tuberosity of the tibia, becomes tendinous, passes towards the metatarsus, and blends with the tendons of the posterior tibial and the long proper flexor of the great-toe. In the dog and the cat it is blended with this latter only.

Tibialis Posticus (Fig. 85, 14; Fig. 86, 12; Fig. 87, 20; Fig. 88, 16).—This muscle arises from the external tuberosity of the tibia, and from the head of the fibula. Thence it passes to the tarsus, and terminates in different fashion in carnivora and other quadrupeds.

In the dog and the cat, it is inserted into the ligamentous apparatus of the tarsus, or into the base of the second metatarsal.

In the other quadrupeds with which we are here occupied it is blended with the long proper flexor of the great-toe.

It is accordingly in the carnivora that the mode of termination of the tibialis posticus most nearly resembles that of this same muscle in the human species. From this independence there results a special action.

It is an adductor and internal rotator of the foot.

Flexor Longus Pollicis (Fig. 84, 18; Fig. 85, 14; Fig. 86, 12; Fig. 87, 18; Fig. 88, 16).—This muscle, as that in man, is the most external of the deep layer of the leg. It is on the external aspect of the latter we perceive it, between the peroneals and the gastrocnemius or tendo-Achillis.

It arises from the fibula and tibia, and is thence directed towards the tarsus. It unites with the long common flexor of the toes to form with it *the deep flexor of the phalanges*, of which it is the principal fasciculus. We may add that in the dog and the cat the posterior tibial remains independent of this latter, but that in the pig, ox, and horse the posterior tibial is united to the preceding to form with them the deep flexor muscle.

Thus constituted, the deep flexor goes towards the phalanges, where it terminates as the deep flexor of the digits of the fore-limbs. In animals possessed of a canon it contributes to form the *tendon* (Fig. 85, 16; Fig. 86, 14, 14; Fig. 88, 19, 19).

Muscles of the Foot

We must remember that on the dorsal surface of the foot in man we find but a single muscle—the dorsalis pedis. The remaining subcutaneous structures of this region consist of the tendons of the anterior muscles of the leg which occupy this dorsal aspect.

Dorsalis Pedis (Fig. 84, 19).—Also called the extensor brevis digitorum, the dorsalis pedis muscle is found in all domestic quadrupeds; but its development is so much the less as the number of digits is more reduced.

In the dog and the cat it arises from the calcaneum, and is inserted into the three internal toes (the first toe excepted) by uniting with the corresponding tendons of the common extensor.

In the pig its disposition is analogous.

As for the dorsalis pedis of the ox and the horse, it is extremely rudimentary, and occupies the superior part of the canon. It is a small, fleshy body, situated on the anterior surface of the metatarsus, which arises from the calcaneum, whence it passes to unite at its inferior extremity with the tendon of the extensor of the phalanges.

As regards the muscles of the sole of the foot, we think it unnecessary to occupy ourselves at any length with them because of their slight importance with regard to external form.

We will only recall that in the median portion of this plantar surface we find in man the short flexor of the toes, which in quadrupeds arises higher up, from the posterior surface of the femur; that it belongs to the muscles of the leg; and that it forms the superficial flexor of the toes, which we have already studied.

We may further add that the suspensory ligament of the fetlock in ruminants and solipeds represents, as in the fore-limbs, the interosseous muscles.

MUSCLES OF THE HEAD

We will divide these muscles into two categories: masticatory and cutaneous.

Masticatory Muscles

The muscles of this group which specially interest us are the masseter and the temporal. As regards the pterygoids, since they are situated within the borders of the inferior maxillary bone, and consequently do not reach the surface, we shall not require to occupy ourselves with them here.

Masseter (Fig. 89, 2; Fig. 90, 1; Figs. 91, 92).—For those who have studied the masseter of man, it is not difficult to recognise that of quadrupeds. Nevertheless, the particular aspect which it presents in different species gives to its study a certain interest.

Arising from the zygomatic arch, and passing downwards and backwards, it is inserted into the external surface of the ramus of the mandible and into its angle.

Its posterior border is in relation with the parotid gland (Fig. 90, 14; Figs. 91, 92), this gland being situated between the corresponding border of the lower jaw bone and the transverse process of the atlas. Such are the general characters; the following are the particular ones:

In the carnivora it is thick and convex. In the horse it is flat, but more expanded; it forms the *flat of the cheek*. In the ox it is flat, as in the latter; but, while being less thick, it is more prolonged in the vertical direction.

The form of the osseous parts which give it origin is, besides, in relation with these differences, and explains the peculiar characters which the masseter presents.

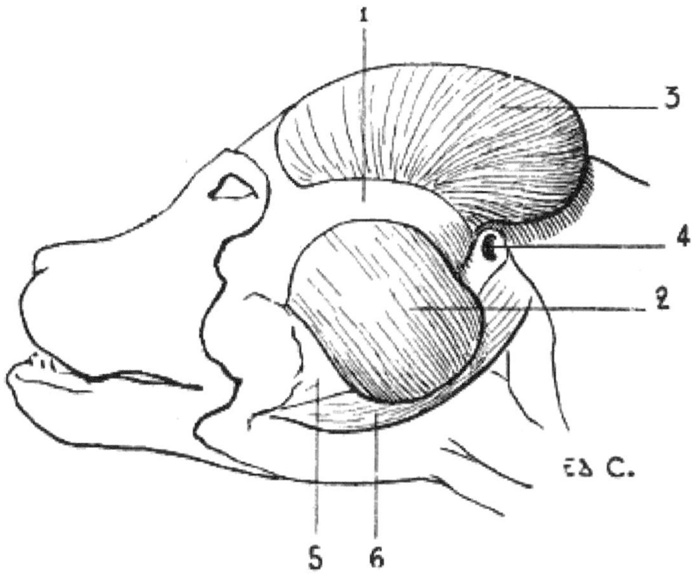

Fig. 89.—Myology of the Dog: Masticatory Muscles (a Deeper Dissection than that shown in Fig. 90).

1, Zygomatic arch; 2, masseter; 3, temporal exposed by the suppression of the auricular and occipital muscles and the pinna of the ear; 4, auditory canal; 5, inferior maxillary bone; 6, digastric.

Indeed, in the dog and the cat the zygomatic arch, strongly convex, springs up in a marked manner from the plane of the lateral aspects of the skull.

In the horse the same arch, less prominent externally, is prolonged by a rectilinear crest on the superior maxillary bone, where it is continued in forming the zygomatic or maxillary spine.

In the ox the same crest ascends a good way towards the inferior margin of the orbit in a curved direction with the concavity inferior, to redescend afterwards on the external surface of the superior maxilla.

The masseter is an elevator of the lower jaw. It acts, above all, as in the human species, in the process of mastication.

Temporal Muscle (Fig. 89, 3).—The development of the temporal is in proportion to the energy of the movements of elevation which the lower jaw has to execute.

It arises from the temporal fossa, and is inserted into the coronoid process of the inferior maxilla.

Its development, enormous in the carnivora, is such that the muscle projects beyond its fossa. It is less voluminous in the horse, and still less so in the ox. In the latter, indeed, the temporal fossa, although deep, is of small extent (see Fig. 62); the frontal bone being large, it is found to be thrown back on the lateral walls of the cranium, below the osseous processes which support the horns and overhang the fossa in question, as well as the muscle which it contains.

It is covered by the auricular muscles, and by the base of the pinna of the ear.

Like the masseter, the temporal is an elevator of the lower jaw.

Cutaneous Muscles of the Head

Occipito-Frontalis.—The epicranial aponeurosis is extremely thin. In the dog the occipital muscle occupies the superior part of the head; it overlies the temporal muscle.

With regard to the frontal muscle, which is of great extent in the ox (Fig. 91, F), it is represented in the horse and the carnivora by a small fleshy fasciculus only, the *fronto-palpebral muscle*, similar to the superciliary muscle. This, occupying the superior and internal part of the border of the orbit, ends by blending its fibres with those of the orbicular muscle of the eyelids at the region of the eyebrow.

Orbicularis Palpebrarum (Fig. 90, 2; Figs. 91, 92).—This annular muscle surrounds the palpebral orifice, and takes its origin on the internal part of the orbital region. In the horse it arises, by a small tendon, from a tubercle which occupies the external surface of the os unguis, or lachrymal bone.

This muscle determines the narrowing and closure of the palpebral orifice.

Pyramidalis Nasi.—The pyramidal muscle is not found in the domestic animals. It appears to be blended with the internal elevator of the upper lip and wing of the nose; this is easy of comprehension if we bear in mind the relative position of these two muscles in the human species.

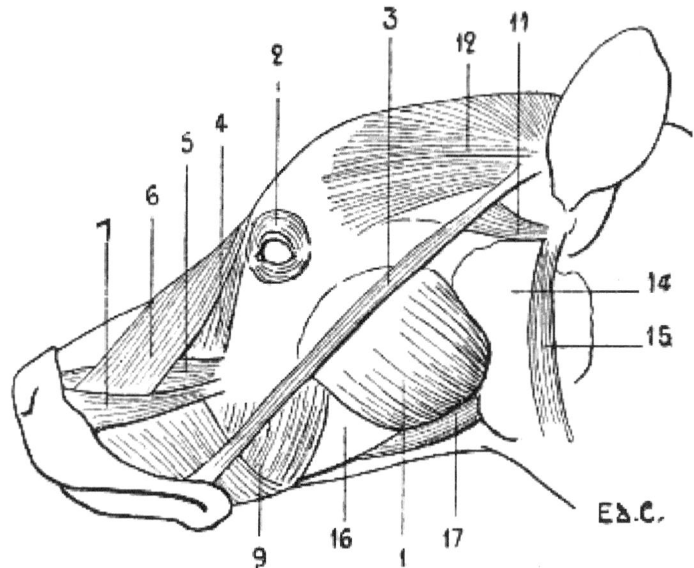

Fig. 90.—Myology of the Dog: Muscles of the Head.

1, Masseter; 2, orbicularis palpebrarum; 3, zygomaticus major; 4, lachrymal (this muscle is sometimes described under the name of the small zygomatic); 5, levator labii superioris proprius; 6, levator labii superioris alæque nasi; 7, caninus; 9, buccinator; 11, zygomatico-auricularis; 12, external temporo-auricularis; 14, parotid gland; 15, parotido-auricularis; 16, inferior maxillary bone; 17, digastric.

Corrugator Supercilii.—This muscle is represented by the fronto-palpebral muscle noticed above, which is by some regarded as a vestige of the frontal.

Zygomaticus Major (Fig. 90, 3; Figs. 91, 92).—This is the *zygomatic-labial* of veterinarians. This muscle is of an elongated form, and has a ribbon-like aspect.

In the dog and the cat it arises from the base of the pinna of the ear, from the portion of this base which bears the name of scutiform cartilage. (With regard to this cartilage, **Zygomatico-auricularis**.) From this it is directed downwards and forwards, to terminate, after having crossed the masseter, on the deep surface of the skin of the corresponding labial commissure.

This mode of termination is the same in the ox and the horse; but where the muscle differs is at the level of its upper extremity. There it ascends less than in the carnivora. In the ox it arises from the zygomatic arch in the neighbourhood of the temporo-maxillary articulation; in the pig and the horse its origin is still lower, on the surface of the masseter, close to the maxillary spine.

When it contracts, it draws upwards the labial commissure.

Now, in man, we remember, it is the great zygomatic that, by an action of the same kind, determines the essential characters of the expression of laughing.

There is, accordingly, a connection to be established between those displacements which are similar and the analogy of facial expression which necessarily results from them.[33]

[33] Édouard Cuyer, 'The Mimic,' Paris, 1802.

Zygomaticus Minor (Fig. 90, 4; Figs. 91, 92).—The existence of this muscle has not been clearly demonstrated. Nevertheless, Straus-Durckheim noted its presence in the horse, and described it as 'a muscle arising by two heads, of which one, the superior, arises from the malar bone below the orbit, and passes downwards and forwards over the fibro-adipose layer which supports the moustache. The second, the inferior, arises from the alveolar border in front of the second molar tooth, and passes forward to be inserted into the same fibro-adipose layer.'[34]

[34] H. Straus-Durckheim, 'Anatomie descriptive et comparative du chat,' Paris, 1845, t. ii., p. 210.

In connection with other quadrupeds, it is described by certain authors as a very thin muscle, arising below the cavity of the orbit, where it is blended with the fibres of the internal elevator of the upper lip and the ala of the nose; thence it proceeds to terminate below by uniting with the subcutaneous muscle. But this muscle is regarded by other authors as the lachrymal muscle, which does not exist in this state in man, but of which the development is particularly remarkable, as to extent, in the ox, in which it descends as far as the buccinator.

According to other authors, some of the fibres of this muscle constitute the small zygomatic.

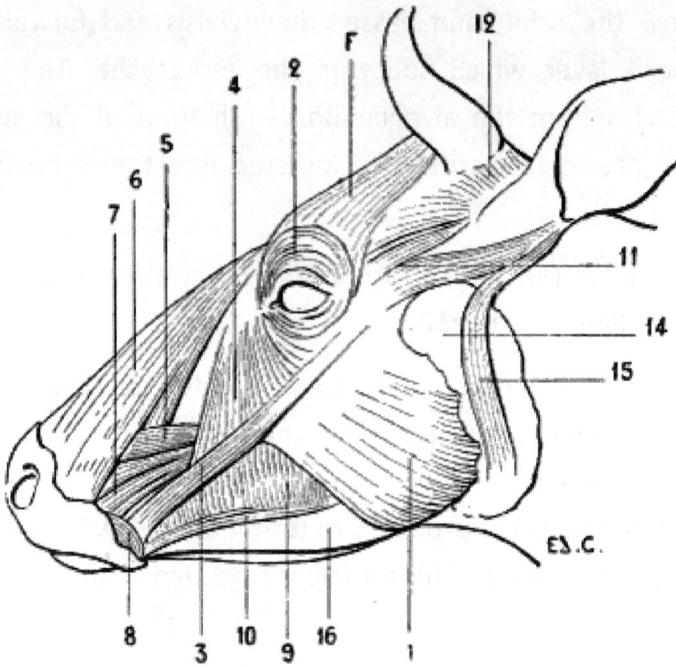

Fig. 91.—Myology of the Ox: Muscles of the Head.

1, Masseter; 2, orbicularis palpebrarum; F, frontalis; 3, zygomaticus major; 4, lachrymal (this muscle is sometimes described under the name of small zygomatic); 5, levator labii superioris proprius; 6, levator labii superioris alæque nasi; 7, levator anguli oris or caninus; 8, orbicularis oris; 9, buccinator; 10, maxillo-labial; 11, zygomatico-auricularis; 12, external temporo-auricularis; 14, parotid gland; 15, parotido-auricularis; 16, inferior maxillary bone.

Levator Labii Superioris Proprius (Fig. 90, 5; Figs. 91, 92).—Also named by veterinarians the *supramaxillo-labial*, or again, the *proper elevator of the upper lip*, this muscle arises from the external surface of the superior maxillary bone, passes under the superficial elevator, which we shall study in the succeeding paragraph, and goes to be inserted into the thickness of the lip, to which its name indicates that it belongs.

The peculiarities of this muscle in different animals are the following:

In the dog and the cat it arises behind the infra-orbital foramen.

In the pig it arises from a depression below the orbital cavity, and its fleshy body is terminated in front by a strong tendon in the upper part of the snout, in which it divides into fasciculi.

In the ox it arises from the maxillary spine.

In the horse it arises below the orbital cavity; then, after having crossed the superficial elevator, it ends in a tendinous expansion, situated in the median line between the nasal fossæ. This expansion divides into fasciculi, which end in the thickness of the upper lip.

By the contraction of this muscle, the lip is raised, on one side only, if a single muscle contracts, or in its whole extent, if the two muscles act simultaneously.

Internal Elevator (or Superficial) of the Upper Lip and the Wing of the Nose (*levator labii superioris alæque nasi*) (Fig. 90, 6; Figs. 91, 92).—This is the muscle veterinarians designate *the supranaso-labial*.

Arising from the frontal and nasal bones, it thence passes towards the upper lip, where it is inserted as well as into the wing of the nose.

In the ox it is united above with the frontal muscle, and below is divided into two fasciculi, between which pass the elevator described above and the canine muscle.

In the horse it is also divided into two fasciculi; but the arrangement is the opposite as regards, their relations with neighbouring muscles, in this animal and in the preceding.

In the ox the external fasciculus is covered by the external elevator and the canine, which pass under the internal fasciculus; in the horse the deep elevator passes under the two fasciculi, and the canine passes under the external bundle, and afterwards covers the internal.

In the pig, the internal elevator is wanting.

As its name indicates, it raises the upper lip and the wing of the nose.

Transversus Nasi.—In the horse this muscle, which is very thin, is situated on the dorsum of the nose, and proceeds to be inserted into the cartilaginous skeleton of the nostrils. In the pig, it occupies an analogous situation. It does not exist in the ox or in carnivora. The transversus nasi is a dilator of the nostrils.

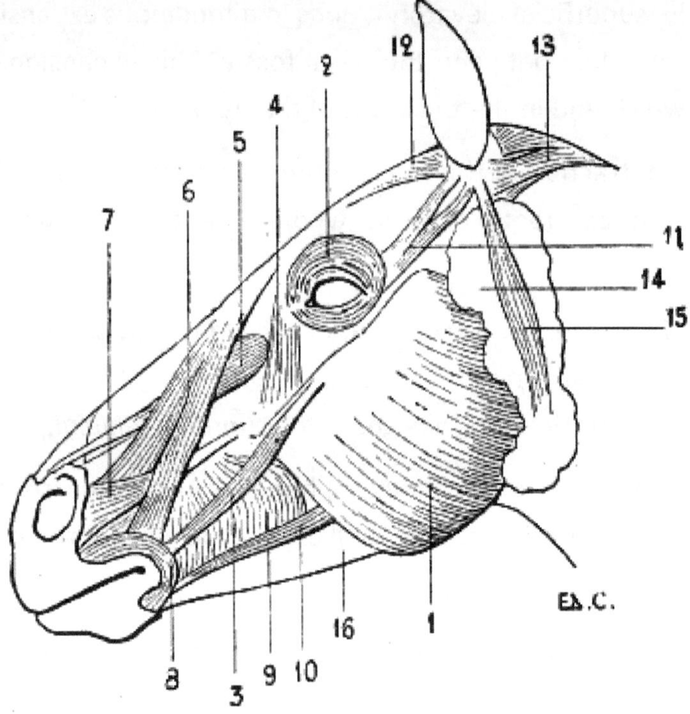

Fig. 92.—Myology of the Horse: Muscles of the Head.

1, Masseter; 2, orbicularis palpebrarum; 3, zygomaticus major; 4, lachrymal (this muscle is sometimes described under the name of the small zygomatic); 5, external elevator (or deep) of the upper lip and ala of the nose; 6, internal elevator (or superficial) of the upper lip and of the ala of the nose; 7, levator anguli oris or caninus; 8, orbicularis oris; 9, buccinator; 10, maxillo-labialis; 11, zygomatico-auricularis; 12, temporo-auricularis externus; 13, cervico-auricularis; 14, parotid gland; 15, parotido-auricularis; 16, inferior maxillary bone.

Caninus (Fig. 90; Figs. 7, 91, 92).—This is the muscle called by veterinarians *the great supramaxillo-nasal.*

In the dog and the cat it is situated below the inferior border of the external elevator of the upper lip, of which it follows the direction. It arises, as does this latter, from the external surface of the maxilla, and goes also to terminate in the upper lip by blending with the internal elevator of this lip and of the alæ of the nose. It raises the upper lip.

In the ox, it arises from the maxillary spine, and then divides into three parts; the superior passes under the internal portion of the internal elevator of the upper lip and the alæ of the nose, and goes into the nostril; whilst the two others, situated lower down, terminate in the upper lip.

In the pig, it is formed of two superimposed fasciculi, which arise from the spine of the maxilla and the impressions in front of it. These two fasciculi terminate in the snout, which they move laterally.

In the horse, it is situated at a certain distance from the external elevator; in the preceding animals it is in contact with the latter. Arising behind from the external surface of the maxilla, in front of the maxillary spine, it is directed towards the anterior part of the face, passes under the external portion of the internal elevator (it is the opposite of this in the ox), and proceeds, on expanding, to terminate in the skin of the nostril. By its contraction it dilates the latter.

Orbicularis Oris (Fig. 91, 8; Fig. 92).—This muscle, very fleshy in the solipeds and the ruminants, is arranged as a ring round the buccal orifice, in the thickness of the lips, where it is blended with the other muscles of this region.

Having for its function the narrowing of the orifice it surrounds, it acts during suction and in the prehension of food.

Triangularis Oris.—This muscle does not exist in domestic quadrupeds.

Quadratus Menti.—In the pig and the carnivora, it arises from the anterior part of the body of the inferior maxillary bone, and passes at the other end to terminate in the corresponding portion of the lower lip, which it depresses by its contraction.

In the ox and the horse this muscle does not exist; it is replaced for the depression of the lower lip, which it affects in other animals, by supplemental fibres of the buccinator.

The Prominence of the Chin.—Below the lower lip in the horse is situated the so-called *prominence of the chin*, limited posteriorly by the *beard*, a depressed region which gives point to the curb of the bridle.

The prominence, which also exists in the ox, is a fibro-muscular pad which blends with the orbicular muscle of the lips, and receives on its superior aspect the insertion of the two muscles (*levator menti*) by which it is suspended. These arise, above, on each side of the symphysis of the inferior maxillary bone. They raise the lower lip with force, and they are the agents which, as we can sometimes observe in the horse, make it click against the upper lip, suddenly projecting it upwards. This action sometimes becomes a habit, and its continuance constitutes a vice.

A corresponding structure is found in the pig and in the carnivora, but in them it does not produce an external prominence such as we have described.

Buccinator (Fig. 90, 9; Figs. 91, 92).—Further designated by the name of *alveolo-labial*, this muscle is situated on the lateral portions of the face, in the thickness of the cheeks. It consists of two layers, one superficial and the other deep.

The deep portion arises from the portion of the alveolar border of the superior maxillary bone which corresponds to the molar teeth, and from the anterior border of the ramus of the mandible. Thence it is directed forwards, passes under the superficial layer, and blends with the fibres of the orbicular muscle of the lips. To this part of the buccinator some authors give the name of molar muscle.

The superficial portion is formed by fibres which pass from the alveolar border of the superior maxillary bone to the corresponding border of the opposite bone. It is very highly developed in the herbivora.

This muscle acts especially during mastication; it serves to press back again under the molar teeth the portions of food which fall outside the dental arch.

In the pig, the ox, and the horse, a muscle which is considered as supplemental to the buccinator is placed along the inferior border of the latter.

This muscle, which we describe separately under the names of *maxillo-labialis* (Fig. 91, 10; Fig. 92) and *depressor of the lower lip*, is clearly distinct from the buccinator, especially in the horse. It arises, behind, with the deep layer of the muscle to which it is annexed, from the anterior border of the ramus of the lower jaw; in front it terminates in the thickness of the lower lip.

In the ox, it is more intimately united with the buccinator.

It depresses the lip to which it is attached, and displaces it laterally when it acts on one side only.

In the human species, the pinna of the ear being generally immobile, the muscles which belong to it are, very naturally, considerably atrophied. Accordingly, the auricular muscles, anterior, superior, and posterior, are reduced to pale and thin fleshy lamellæ, whose action is revealed in certain individuals, only in a way which may be said to be abnormal.

It is not the same in quadrupeds. The pinna of the ear is extremely mobile, and its displacements have a real value from the point of view of physiognomical expression. It is therefore necessary to review the muscles which move this pinna without giving them, at the same time, more importance than they merit, since in themselves they do not determine the formation of surface reliefs, which are sufficiently apparent.

Notwithstanding that for certain of these muscles it is possible to trace their analogy with those of the auricular region of man, it is very difficult, because of their complexity, to trace this analogy for all. This is why we shall not be able here, as we have done for the other muscles of the subcutaneous layer, to give at the head of each paragraph the name of a human muscle, and then to group in the same paragraph the muscles which correspond to it in different quadrupeds. Therefore the nomenclature and the divisions adopted for these latter must serve us as a base or starting-point.

Because the pinna of the horse's ear is so very mobile, we will first begin with a study of its auricular muscles.

Zygomatico-auricularis (Fig. 92, 11).—This muscle, which is formed of two small bands of fleshy fibres, arises from the zygomatic arch in blending with the orbicular muscle of the eyelids; thence it is directed towards the base of the pinna of the ear, and is inserted into this base, and also into the cartilaginous plate situated in front of and internal to this, and resting on the surface of the temporal muscle; this is the scutiform cartilage.

The zygomatico-auricularis, which we look on as the homologue of the anterior auricular of man, draws the pinna of the ear forwards.

Temporo-auricularis Externus (Fig. 92, 12).—This, which is thin and very broad, covers the temporal muscle.

It arises from the whole extent of the parietal crest, blending in this plane, in its posterior half, with the muscle of the opposite side. Thence it is directed outwards towards the pinna of the ear, and is inserted into the internal border of the scutiform cartilage and on the inner side of the concha—that is to say, of the conchinian cartilage—which forms the principal part of the pinna. We are supposing, in the description of the muscles which move it, that this pinna has its opening directed outwards.

The external temporo-auricular, which recalls, from its situation, the superior auricular of man, is an adductor of the ear; besides, it causes it to describe a movement of rotation from without inwards, so as to direct its opening forwards.

Scuto-auricularis Externus.—This muscle may be considered as supplementary to the external temporo-auricular; the concha fasciculus of this latter partly covers it.

Extending from the scutiform cartilage to the inner side of the concha, it contributes to the movement of rotation by which the opening of the pinna of the ear is directed forwards.

Cervico-auricular Muscles (Fig. 92, 13).—These muscles, three in number, are situated behind the pinna of the ear; they are called, from their mode of superposition, the superficial, middle, and deep.

These arise, all three, from the superior cervical ligament, and pass from there towards the cartilage of the concha. They recall, as regards situation, the posterior auricular muscle of man.

Superficial Cervico-auricular (*Cervico-auricularis superioris*).—This muscle, inserted into the posterior surface of the concha, draws this cartilage backwards and downwards.

Middle Cervico-auricular (*Cervico-auricularis medius*).—Situated between the two other muscles of the same group, it proceeds, after having covered the superior extremity of the parotid gland, to be inserted into the external part of the base of the concha. It determines the rotation of this concha in such a way as to direct the opening of the ear backwards.

Deep Cervico-auricular (*Cervico-auricularis inferioris*).—Covered by the preceding muscle and the superior portion of the parotid, it is inserted into the base of the pinna of the ear, and has the same action as the middle cervico-auricular.

Parotido-auricularis (Fig. 92, 15).—This is a long and thin fleshy band which arises from the external surface of the parotid gland, and tapering as it passes upwards towards the pinna of the ear, is inserted into the external surface of the base of the concha, below the inferior part of the angle of reunion of the two borders which limit its opening.

It inclines the pinna outwards; it is, accordingly, an abductor of the pinna.

Temporo-auricularis Internus.—This muscle is covered by the external temporo-auricular and the superior cervico-auricular. It arises from the parietal crest, and is inserted into the internal surface of the concha. It is an adductor of the pinna of the ear.

There are, finally, an internal scuto-auricular muscle and a tympano-auricular; but they do not present any interest for us; we can simply confine ourselves to making mention of them.

In the ox, because of the situation of the temporal fossa and the fact that the external temporo-auricular muscle is applied, as in the horse, over the muscle which this fossa contains, this temporo-auricular muscle does not reach the middle line (Fig. 91, 12).

But in the cat and the dog this muscle covers all the upper part of the head (Fig. 90, 12). It is divided into two parts: the interscutellar and the fronto-scutellar.

The interscutellar is a single muscle, thin and broad, covering the temporal muscle and a portion of the occipital, extending from the scutiform cartilage of the pinna of one side to the same cartilage of the pinna belonging to the side opposite. It approximates the two pinnæ to one another by bringing them each into the position of adduction.

The fronto-scutellar arises from the orbital process of the frontal bone, and from the orbital ligament, which at this level completes the interrupted osseous boundary of the orbital cavity. Thence it is directed, widening as it proceeds, towards the scutiform cartilage, and is there inserted by blending with the corresponding part of the great zygomatic. Its action is analogous to that of the preceding muscle; but, further, it directs the opening of the pinna forwards.

These are the muscles which act, for example, when the dog, having his attention strongly attracted by any cause, pricks up his ears and turns the openings forward, in order the better to understand every sound which proceeds, or may possibly proceed, from that which he observes. From this, which may be extremely well seen in some individuals, results the appearance of vertical wrinkles of the skin in the interval between the pinnæ of the ears, these being caused by the folding of the integument, whilst the pinnæ approach one another. These movements, with which are associated fixation of look and a widening of the palpebral fissure, produce a peculiarly expressive look; this is why they merit our attention.

Zygomatico-auricularis (Fig. 90, 11).—Arises from the internal surface of the great zygomatic, passes towards the pinna of the ear, and goes to be inserted into the external part of the base of the pinna, below its opening, to a prominence which corresponds to the antitragus of the human ear. It is to this antitragus, but proceeding from another direction, that the parotido-auricular muscle is inserted (Fig. 90, 15).

With regard to the cervico-auriculars, they are all three present. The superior, or superficial, situated behind the interscutellar portion of the external temporo-auricular, has its origin on the median line of the neck; thence it passes towards the pinna of the ear, blending its fibres with those of the interscutellar muscle, and is inserted into the scutiform cartilage and the internal surface of the pinna.

Such are the principal muscles of the ear in the carnivora; it would seem to us superfluous to dwell on the others of this region, so that we will here conclude the study of the muscles in general, and that of the myology of the head in particular.

CHAPTER III

EPIDERMIC PRODUCTS OF THE TERMINAL EXTREMITIES OF THE FORE AND HIND LIMBS

We will first recall to mind that among the quadrupeds some are found of which the fingers and toes have their third phalanges terminated by claws—these are the unguiculates; and that in others the terminal extremity of each limb is completely encased in a horny envelope, the hoof—these are the ungulates.

In the first group, the claws remind us to a certain extent of the arrangement of the nails in man; the inferior aspect of the paws is covered by an epidermic layer, thick and protective, which may be likened to the skin, correspondingly thick, which covers in the greater part of its extent the plantar surface of the foot in the human species.

In the second group, the surface by which the third phalanx rests on the ground is correspondingly protected, but this time by a layer of horn which belongs to the hoof.

After the preceding remarks, our study will be found to fall into a natural division, and it is in the order which we have just followed for the purpose of indicating its existence that we now proceed to study the nature and form of the different elements which complete or protect the digital extremities of the thoracic and abdominal limbs.

Claws.—These horny coverings of the third phalanges, which we have to consider only in the dog and cat, may be compared with the nails of man, with which, however, they present, as is well understood, characteristic differences.

The claws are compressed laterally, curved on themselves, and are terminated in front by a sharp point in the felide, but more blunted in the dog. Their superior border is convex and thick. We may say, therefore, that a claw is a sort of hollow tube, in the form of a cone flattened in the transverse direction, in which the third phalanx is set, and which is itself set in a groove formed by a kind of osseous hood which occupies the base of this third phalanx (see Fig. 37).

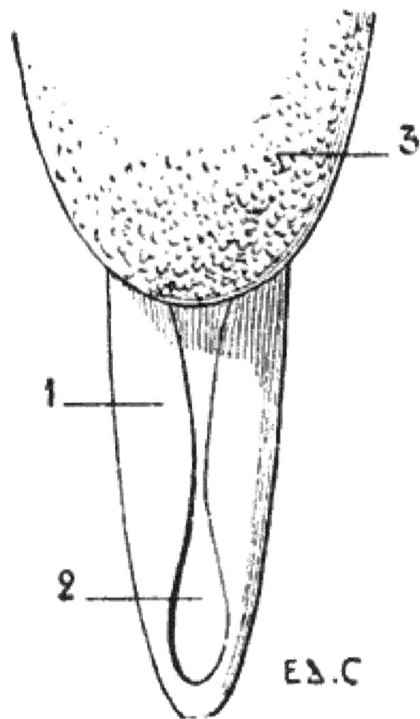

Fig. 93.—Claw of the Dog: Inferior Surface.

1, Horny lamina of the claw; 2, plantar nail; 3, tubercle of the corresponding digit.

This definition is exact, as regards the general appearance; but, when more closely scrutinized, it is not sufficient. The tube in question is not formed of a single piece; each of the claws is formed by a lamina laterally folded, but of which the borders are not exactly joined together inferiorly; they leave between them a small interval, and this is filled by a layer of more friable horny substance, to which has been given the name of plantar nail. This arrangement, which is clearly defined in the dog (Fig. 93), is comparable to that which we shall afterwards meet with in connection with the sole of the hoof of the horse (see Fig. 100). In the dog and the cat, the weight of the limb resting on the inferior surface of the phalanges, it was necessary that the region of the plantar surface of the foot corresponding to these latter should be protected; this is the function of certain fibro-adipose pads, which are situated there, and which are designated by the name of *plantar tubercles*.

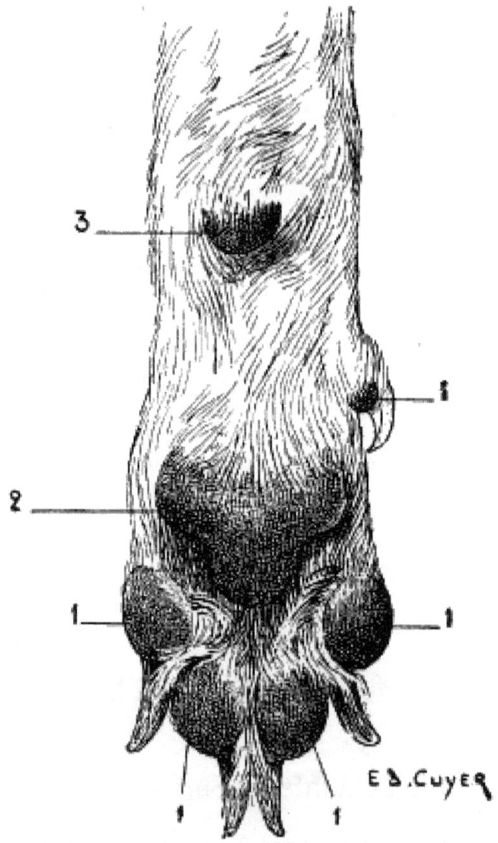

Fig. 94.—Left Hand of the Dog: Inferior Surface, Plantar Tubercles.

1, 1, 1, 1, 1, Tubercles of the fingers; 2, plantar tubercle; 3, tubercle of the carpus.

Plantar Tubercles (Fig. 94).—These tubercles, or dermic cushions, are divided, in each paw, into *tubercles of the digits* (or of the toes), a *plantar tubercle*, and, on the fore-limbs, a *tubercle of the carpus*.

The tubercles of the fingers (or of the toes) are of the same number as the latter. That which belongs to the thumb is but little developed, but the others are more so. They are in relation with the plantar surfaces of the second and third phalanges, so that when the paw is in contact with the ground the articulation which, in each of the fingers or toes, joins these phalanges, rests on the corresponding pad.

The plantar tubercle is larger than the preceding. It is of a more or less rounded form; sometimes it is triangular, and then comparable in outline to the ace of hearts, the point of which is, in this case, turned towards the claws; its margin being sometimes strongly indented, it may also have a trilobate form. It is on it that rest the metacarpo-phalangeal or metatarso-phalangeal articulations, according to the limb studied. The tubercle of the carpus, situated at the level of the posterior surface of this latter, is less important than the preceding, the region which it occupies not reaching the ground during walking. But it is not to be neglected from the point of view of external form, because of the prominence which it produces.

In the ungulates the terminal extremity of the limb is, as we have above pointed out, enclosed in a horny envelope which is no other than the hoof.

We will first study the hoof of the horse—a hoof which is single for each of the limbs, inasmuch as in this animal each of these has but a single digit.

Hoofs of the Solipeds.—We will first study the hoof as regarded in a general way—that is, without taking into account the limb to which it belongs. We will afterwards point out the differences presented when the hoofs of the fore and hind limbs are compared.

In connection with the external forms of the horse, the study which we are now commencing is of great importance. But, before entering upon it, it appears to us necessary to rapidly examine what the hoof contains (Fig. 95).

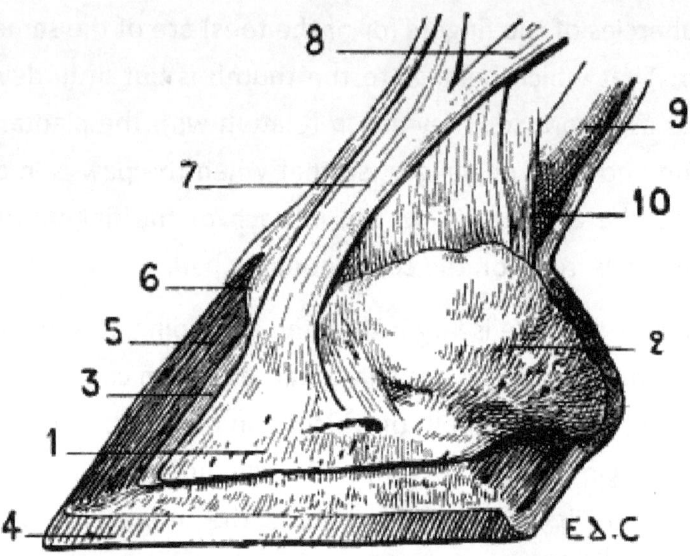

Fig. 95.—Vertical Antero-posterior Section of the Foot of a Horse.

1, Third phalanx; 2, fibro-cartilage; 3, podophyllous tissue; 4, inferior part of the wall; 5, section of the wall of the hoof; 6, cutigerous cavity; 7, tendon of the anterior extensor of the phalanges; 8, reinforcing band coming from the suspensory ligament of the fetlock; 9, tendon of the superficial flexor of the phalanges; 10, tendon of the deep flexor of the phalanges.

In the interior of this horny box we find the third phalanx, a small sesamoid bone placed opposite to the posterior border of the latter, a portion of the inferior extremity of the second phalanx, and the tendons, which terminate at this region.

To the third phalanx are added two fibro-cartilaginous plates, flattened laterally, which prolong backwards the bone to which they are annexed. The inferior border of each of these fibro-cartilages is fixed by its anterior part to two osseous prominences situated at each of the angles which terminate the small phalanx behind; these prominences are: *the basilar process* and *the retrorsal process* (Fig. 96); by its posterior part, this border is continuous with a structure known as *the plantar cushion* (see further on).

The posterior border is directed obliquely upwards and forwards. The superior border, which is convex or rectilinear, is thin, and is separated from the posterior border by an obtuse angle. Finally, the anterior border, which is directed obliquely downwards and backwards, is united to the ligamentous apparatus, which keeps the second and third phalanges in contact.

These fibro-cartilages, at their upper extremities, project beyond the hoof, and therefore assist in the formation of the lateral regions of the foot,[35] at the part which is called the *crown*. They project less above the hoof in the posterior limbs.

[35] Here, for the first time, apropos of the hoof, we use the word 'foot.' As in osteology and in myology we have, for the sake of clearness of comparison, designated under this name the region limited above by the tarsus, it is necessary to point out here that we employ the same word for a more restricted region. This we did in conformity with the usage of veterinarians, who so designate the region of the hoof. It is necessary to explain this double employment of the word, and, further, to show the particular meaning ascribed to it.

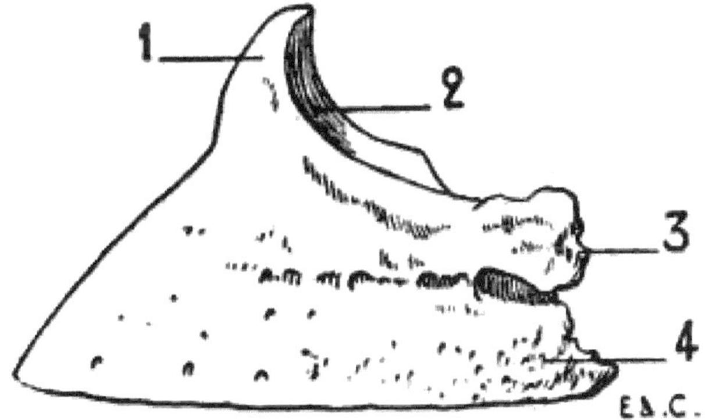

Fig. 96.—Third Phalanx of the Horse: Left Anterior Limb, External Surface.

1, Pyramidal eminence; 2, surface, for articulation with the inferior extremity of the second phalanx; 3, basilar process; 4, retrorsal process.

The posterior and inferior borders of these cartilages meet at an acute angle. The angle so formed, or cartilaginous bulb, constitutes the base of the region, which is commonly called the *heel*—a part of the foot which, as its name implies, is situated posteriorly, but which we must not confound, as we might be led to do, with the region occupied by the calcaneum. We know from our previous studies of comparative osteology that this latter is situated much higher up.

The *plantar cushion* is a sort of fibrous wedge which occupies the interval bounded by the fibro-cartilaginous plates which we have just been studying. Its inferior surface, the form of which we shall find to be reproduced by a portion of the corresponding surface of the hoof, is prolonged anteriorly into a point, while behind it is divided into two branches, which, diverging from one another, join the posterior angles of the fibro-cartilages. These two branches are separated by a median excavation.

The different constituent elements which we have just been discussing give elasticity to the foot.

To finish the examination of the parts contained in the hoof, we will add that among them is also found what is called the fleshy *envelope*, or *flesh* of the foot.

We divide the latter into three regions: the podophyllous tissue, striated or laminated flesh which is spread out over the anterior surface of the third phalanx; the pad, or the hardened skin which corresponds to the upper border of the hoof, and forms a prominence above the podophyllous tissue; and the villous flesh, or velvety tissue which covers the plantar surface of the third phalanx and the plantar cushion. These three tissues form as a whole the keratogenic membrane—that is to say, that which produces horny tissue, and consequently regenerates the hoof.

It is this latter that we now proceed to study.

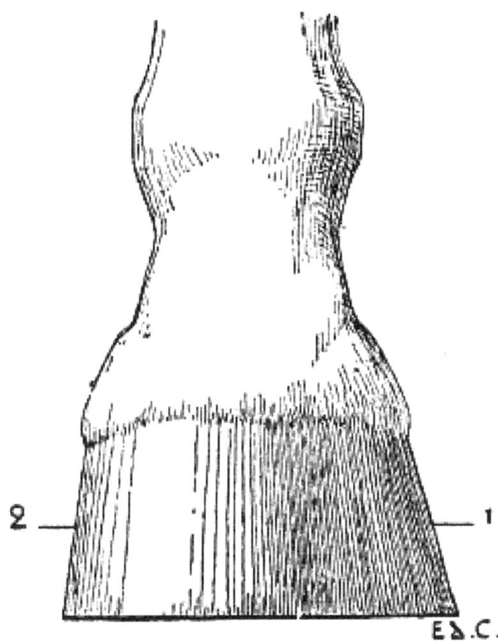

Fig. 97.—Left Anterior Foot of the Horse: Anterior Aspect.

1, Outer side; 2, inner side.

When we examine its anterior surface or the opposite one, the hoof of the horse has the shape of a truncated cone with the base below and the summit cut off obliquely downwards and backwards (Fig. 97).

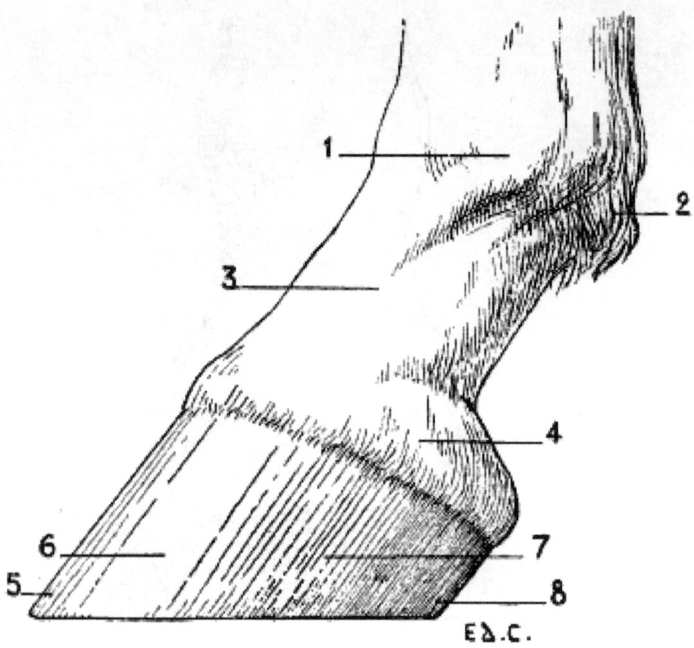

Fig. 98.—Left Anterior Foot of the Horse: External Aspect.

1, Fetlock; 2, spur or beard; 3, pastern; 4, outline determined by the external fibro-cartilage; 5, acute angle; 6, nipple; 7, quarter; 8, heel.

Viewed on one of its lateral aspects, it may be compared to a truncated cylinder placed on the surface of the section (Fig. 98). We particularly call attention to this latter comparison, for it singularly aids us in making a representation of the foot of the horse when viewed laterally.

Notwithstanding that the hoof forms apparently a homogeneous whole, it consists of three parts, which may be separated from one another by maceration. The indication of such disunion, artificially produced, may seem useless. It is not so, however, for this division of the hoof will permit us to carry out the study of the latter in a clearer, and consequently a more satisfactory, way. The three parts in question are the *wall*, or *crust*, the *sole*, and the *frog*.

The *wall* is that portion of the hoof which we see when the foot rests on the ground. It is a plate of horn which, applied to the anterior and lateral surfaces of the foot, diminishes in height as it approaches the posterior part of the region. Posteriorly and at each side the wall is folded on itself, and is then directed forwards to terminate in a point, after having enclosed the frog, which we will soon study.

Although the wall forms a continuous whole, it has been divided into regions to which special names are given. The anterior part, from the superior border to the inferior, is called the *pince* or *toe* for a width of 4 to 5 centimetres. External to the toe, and on each side of it, for a distance of 3 or 4 centimetres, is the *nipple*. Behind the *nipples* are the *quarters*. Still further back, where the wall folds on itself, forming the *buttress*, is found the region of the *heels*. Finally, the portions of the wall which form its continuation in passing forward are called the *bars*.[36] These are only visible on the inferior surface of the hoof (see Fig. 100).

[36] It is to the angle of inflexion or heel that some authors give the name of buttress; it is the bars which other authors designate in this fashion. The wall, convex transversely, is, in its anterior part (viz., the *toe*) inclined strongly downwards and forwards. This obliquity tends to become gradually effaced on the lateral parts to such a degree that at the quarters it becomes almost perpendicular to the surface of the ground. The internal quarter is less rounded than the external; in addition to this (Fig. 97), it approaches more nearly to the vertical direction.

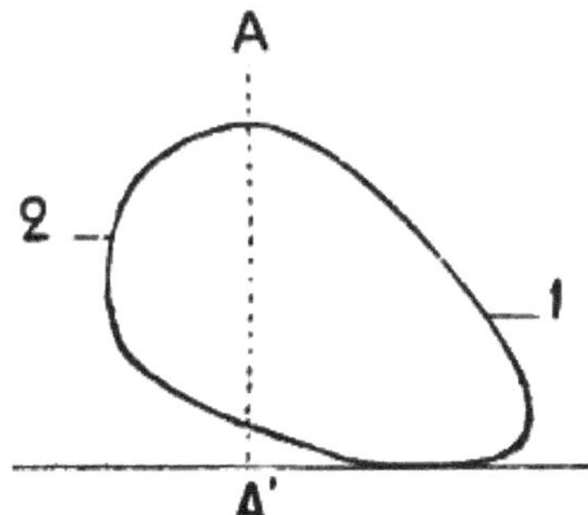

Fig. 99.—Vertical and Transverse Section of a Left Human Foot: Outline of the Divided Surface of the Posterior Segment of this Section (Diagrammatic Figure).

AA′, Vertical axis passing through the middle of the leg and the second toe; 1, outer side; 2, inner side.

In our opinion, this latter difference clearly recalls certain characters of the general form of the human foot. In fact, the latter has its dorsal surface inclined downwards and outwards, whereas its internal border may be said rather to be vertical. A transverse section of the foot (Fig. 99) justifies this comparison, which to us appears interesting, not only as regards the resemblance which exists between these organs of support, but, further, because it constitutes a mnemonic which enables us, on condition that we remember the form of the human foot, to recall the above-described character of that of the horse.

The greater convexity of the outer portion of the hoof is found equally on the human foot; the external border of this foot is more convex than the opposite one.

The inferior border of the wall (Fig. 100) is, in the case of unshod horses, always in wear when in contact with the ground. It is intimately united to the circumference of the sole (see further on).

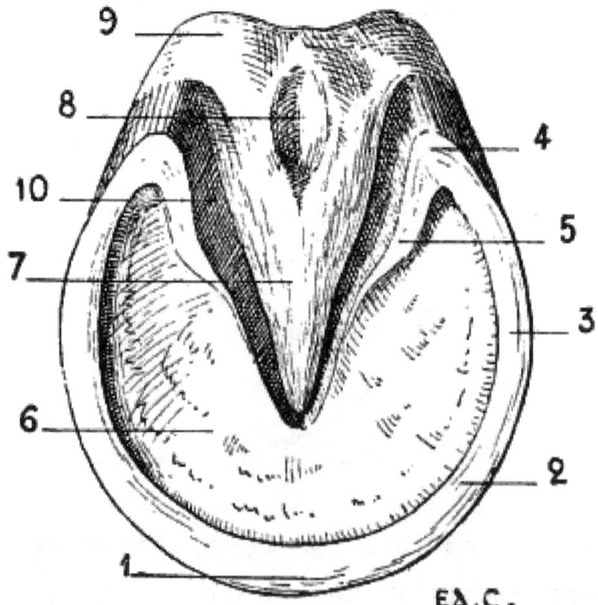

Fig. 100.—Inferior Surface of a Fore-hoof of the Horse: Left Side.

1, Internal border of the wall (toe); 2, wall; 3, quarter; 4, heel; 5, bar; 6, sole; 7, frog; 8, median cavity; 9, prominence of the frog; 10, lateral cavity.

The superior border is hollowed on its internal surface by a groove, the cutigerous cavity or basil, which lodges the cushion (see Fig. 95). We have described this latter above, in connection with the flesh of the foot.

The substance of the wall presents a fibrous appearance which is pretty strongly pronounced. The constituent fibres from which this appearance results are directed from the superior border towards the inferior in parallel and regular lines.

The *sole* is a horny plate which occupies the inferior surface of the hoof (Fig. 100). It is situated between the inferior border of the wall and the bars; and, on account of the oblique direction of these latter, it presents a strongly-marked groove of a **V**-form, with the opening directed backwards. In this depression is lodged the frog.

The inferior surface is concave, and thus forms a sort of vault, more or less deep, according to the individual. The sole has a scaly, laminated aspect.

We have seen (Fig. 93) that on the inferior surface of the claws of carnivora is found a small interval which is filled by a plate of a more friable horny substance, to which has been given the name of the plantar nail. It seems to us that there is an interesting relationship between the said plantar nail and the sole which we have just been studying.

Indeed, these two horny structures appear to be homologous. Is not the lamina of the claw comparable to the wall of the hoof? And does not the interval which occurs at the inferior part of this latter, and is filled by the sole, recall that which, in extremely reduced form, is filled by the plantar portion of the claws?

The *frog* (Fig. 100) is a mass of horn, in form of a wedge, with its apex in front, which occupies the space limited laterally by the recurved portions of the wall (the bars) and the posterior border of the sole.

It covers the plantar cushion previously described and reproduces its form.

Its inferior surface is hollowed out in the middle by an excavation, which is known as the *median lacuna*. This cavity separates the branches of the frog, which terminate posteriorly by two swellings which are known as *the prominences of the frog*, forming two rounded elevations situated above the claws. These same branches unite in front of the median lacuna to form the body of the frog. This latter, in its anterior part, gradually narrows, and terminates in a point which occupies the bottom of the hollow limited laterally by the bars of the wall and the posterior border of the sole.

Between the lateral surfaces of the frog and the bars are found two angular cavities—*the lateral lacunæ*, or the *commissures of the frog*.

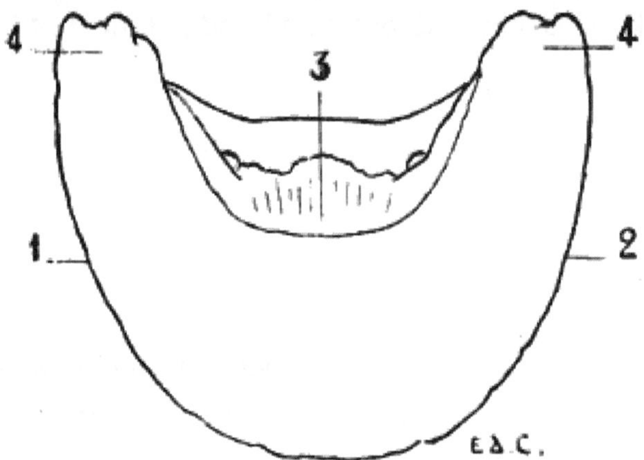

Fig. 101.—Third Phalanx of the Horse: Left Anterior Limb, Inferior View.

1, External border; 2, internal border; 3, semilunar crest; 4, 4, re-entrant processes.

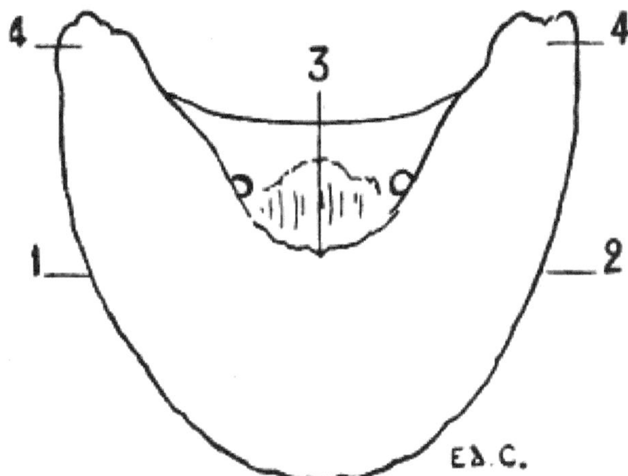

Fig. 102.—Third Phalanx of the Horse: Left Posterior Limb, Inferior View.

1, External border; 2, internal border; 3, semilunar crest; 4, 4, re-entrant processes.

As an indispensable complement to the study which we have just made, it is necessary to add that the hoofs of the fore-limbs and those of the hind ones present differences of form which cannot be ignored—differences which we are already able to conjecture by looking at the respective third phalanges which terminate those limbs, and especially at their inferior surfaces (Figs. 101, 102).

The hoofs of the fore-limbs (see Fig. 100), viewed on their plantar surface, are more rounded than those of the hind-limbs (Fig. 103)—so that their external contour may be compared to a semicircle—whilst the hind-hoofs, which are narrow and of more oval shape, rather recall by their form the aspect of an ogive.

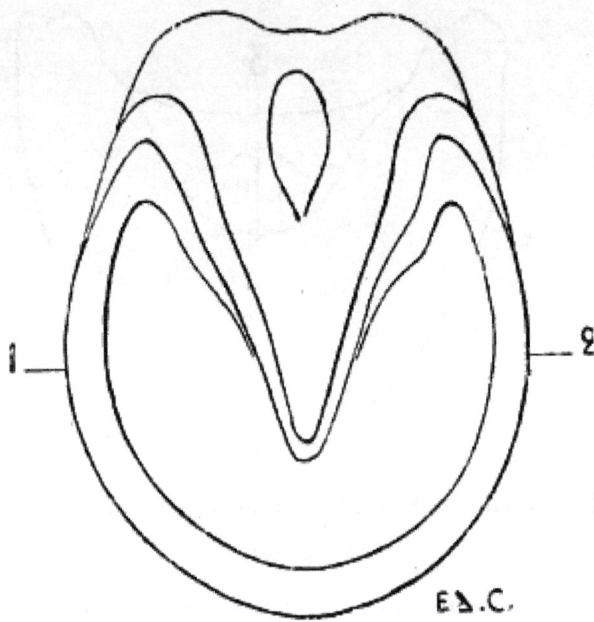

Fig. 103.—Inferior Surface of a Hind-hoof of a Horse: Left Side.

1, External border; 2, internal border.

This seems to result from the fact that the fore-limbs support the more considerable part of the weight of the animal. The best proof which can be given of this overweighting is the eagerness with which very often, when a horse is stopped near the edge of a footpath, for example, he places his fore-feet on the latter. In thus raising his fore-quarters, he throws part of his weight backwards, and in this way relieves his fore-limbs.

With regard to the difference of form which we have just pointed out, we have sometimes heard the following comparison made: the contour of the hoofs of the fore-limbs, viewed from below, recalls that of an apple; that of the hoofs of the hind-limbs recalls the outline of a pear.

As a mnemonic this comparison is insufficient, for nothing connects either of the forms indicated with the region to which the hoofs belong.

We much prefer one made for us this very year by one of the students of our course at the School of Fine Arts, after the lecture in which we had just pointed out the differences in question. Giving the idea of a semicircle and an ogive, which we described above, he remarked to us that the idea would perhaps be more easily fixed in the memory if we associated with it the idea of the chronological order in which the Roman and ogival art succeeded. Indeed, as the Roman art preceded the ogival art, so the hoofs which have the semicircular form precede those which have the form of an ogive.

This interpretation appeared to us ingenious; this is why we wished to give it here a place which seems to us to be merited.

Fig. 104.—Left Posterior Foot of a Horse: External Aspect.

The wall of the hoof of a fore-limb, viewed on one of its lateral surfaces (see Fig. 98), is more oblique than that of one of the hind-hoofs looked at in the same way (Fig. 104). This difference, very marked especially at the region of the toe, is correlated with that of the direction of the pastern. In fact, in the anterior limbs this is a little more oblique than in the opposite ones.

We have still to describe, in connection with the horse, some epidermic tissues, which are known as *chestnuts*.

The chestnut is a small, horny plate which is found on the internal surface of each of the limbs, at a level differing on the anterior from that of the posterior ones.

On the anterior limbs the chestnut is situated on the internal surface of the forearm, towards the middle part, or the inferior third of this region. On the posterior limbs it is developed on the back of the superior extremity of the internal surface of the canon, towards the inferior part of the ham—that is, the tarsus.

Inasmuch as some authors consider the chestnuts as being vestiges of the thumb and the great-toe, we propose giving a mnemonic which will enable us to remember their situation, or, rather, their difference of level.

If we consider that the thumb, in the human species, is longer than the first toe, we may easily remember that the chestnut is placed higher in the anterior limbs than in the posterior ones. Indeed, if we suppose a digit taking its origin at these points, it will be longer in front (the thumb) than behind (the first toe).

Hoofs of the Ox and the Pig.—The ox has four hoofs on each foot—two which contain the third phalanges, and two others, rudimentary, situated at the posterior aspect of the limb, at the level of the inferior part of the canon; these latter bear the name of *spurs*. We will occupy ourselves especially with the former (Fig. 105).

Each of the hoofs presents three faces which, if we consider them in relation to the median axis of the limb to which they belong, are: external, internal, and inferior. The external surface resembles the wall of the hoof of the horse. The internal surface is slightly concave from before backwards, so that the external and internal hoofs of the same foot are not in contact with each other, except by the extremities of this surface, and that an interval separates them between these two points. The inferior surface, slightly depressed, ends behind in a swelling produced by the plantar cushion, which covers a thin lamina of horn.

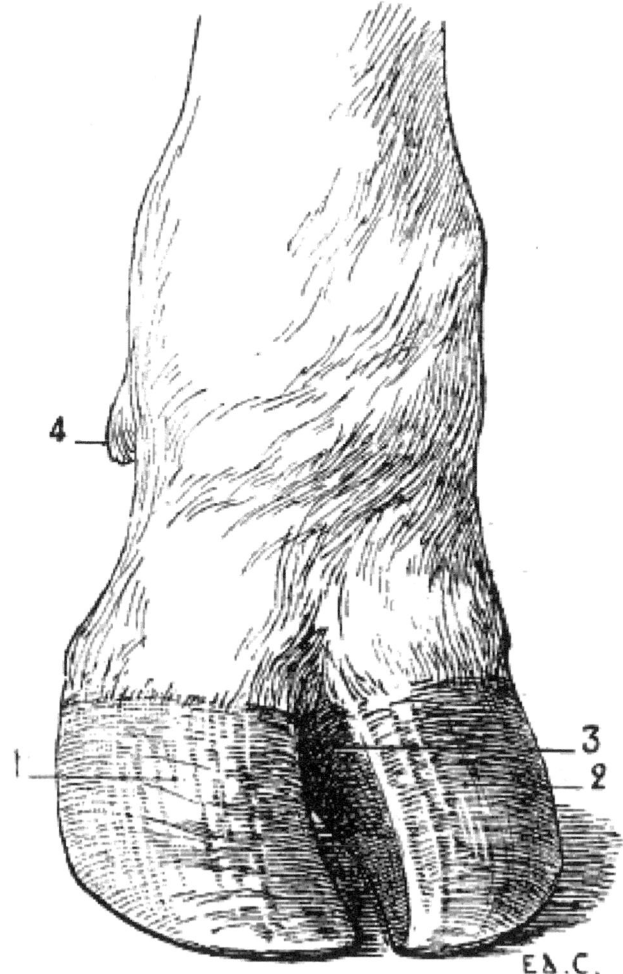

Fig. 105.—Foot of the Ox: Left Side, Antero-external View.

1, Internal hoof; 2, external hoof; 3, internal surface of this latter; 4, internal spur.

At the anterior part of the hoof these three surfaces unite in forming a well-marked angle which, on account of the concavity of the internal surface, is slightly curved towards the axis of the foot.

The pig has also four hoofs—two for the great digits and two for the lateral digits. They recall those of the ox.

CHAPTER IV

PROPORTIONS

Inasmuch as we have taken for granted, in connection with the present volume, that before entering on the study of the anatomy of quadrupeds the reader was prepared for it by a sufficient knowledge of human anatomy, it is quite natural that we should extend the same supposition to the study of proportions.

For this reason, the definition of proportions, considered from a general point of view, their signification, their function and their utility, are questions which it would be superfluous to enter upon here. We will content ourselves by calling to mind that the common measure chosen by preference is the length of the head, and that, ordinarily, it is with it that we compare the dimensions of other parts.

Among the animals whose structure we have examined, there is one of which the proportions deserve to be marked in preference to every other: this is the horse.

Wherefore this preference? In the first place, it is because of the overwhelming position which this animal occupies in the artistic representation of quadrupeds; that it is more frequently associated with man; that, notwithstanding its division into different races, its general proportions may be referred to a special type.

It is also because the indications relative to these proportions will suffice to show the way which the artist must follow in order to find for himself, at the time when the necessity for it arises, the proportions which characterize the other animals.

Our intention is not, in connection with the subject which now occupies us, to enter into a deep discussion on the various opinions which have been set forth. We desire, above all, to give some indications which, from the practical point of view, can be utilized in the representation of the horse, and at the beginning to demonstrate the advantages of these indications. Now, there is a fact which we have had occasion to note; it is the following: almost invariably, when a person who is little accustomed to represent the horse, or not previously informed of certain proportions of lengths, begins to draw from nature, the error generally committed is that of making the head too small and the body too long. Is it a preconceived idea which is the cause that one regards them in this manner? Perhaps. At all events, certain artists who have made the representation of horses their special study have even had this habit. It is therefore necessary to be informed of the proportions; this is the object of the study which we are now undertaking.

Bourgelat,[37] in the eighteenth century, fixed for the first time and in complete fashion the proportions of the horse; it is he, consequently, who created the æsthetics of the horse. It is but justice to recall the fact. His system has a point of analogy with that which is employed to determine the human proportions. Indeed, Bourgelat chose the length of the head as a standard of measurement, and the subdivisions of the head for measures of less extent. 'Since beauty,' said he,[38] 'resides in the congruity and proportion of the parts, it is absolutely necessary to observe the dimensions, individual and relative, and in order to acquire a knowledge of the proportions, to assume a kind of measure which can be indiscriminately common for all horses. The part which can serve as a standard of proportion for all the others is the head. Take a measurement between two parallel lines—one tangent to the nape of the neck or the summit of the forelock, the other tangent to the extremity of the anterior lip—a line perpendicular to these two tangents will give you its geometrical length. Divide this length into three portions, and give to these three parts a special name, which may be applied indefinitely to all heads—as, for example, that of *prime*. Any head whatsoever will, accordingly, in its geometrical length, always have three *primes*; but all the parts which you will have to consider, whether in their length, in their height, or in their width, cannot constantly have either one prime, or a prime and a half, or three primes; subdivide, then, each *prime* into three equal parts, which you will name *seconds*, and as this subdivision will not suffice to give you a just measure of all the parts, subdivide anew each *second* into twenty-four *points*, so that a head divided into three *primes* will have, by the second division, nine *seconds*, and two hundred and sixteen *points* by the last.'

[37] Claude Bourgelat, founder of the veterinary schools in France. He was born at Lyons in 1712, and died at Paris in 1779.

[38] Bourgelat, 'Éléments de l'art vétérinaire. Traité de la conformation extérieure du cheval,' Paris, edition of 1785, p. 133.

But where this system appears to us to have lost somewhat of its unity is when the author transforms it, in pointing out the following mode of procedure: 'But the head itself may err by default of proportion. This part is not, indeed, considered as either too short or too long, too thin or too thick, but by comparison with the body of the animal. Now, the body, being required to have—whether in length, reckoning from the point of the arm to the prominence of the buttock, or in height, reckoning from the summit of the withers to the ground—two heads and a half; whenever the head, by its geometrical length, shall give, in length or in height, to the body measured more than two and a half times its own length, it will be too short; and if it gives less, it will be too long.

'In the case in which one of these faults exists there would be no further question of establishing by its geometrical length the proportions of the other parts. Give up this common measure, and measure the height or the length of the body; divide the length or the height into five equal portions; take, then, two of these divisions, divide them into *primes*, *seconds*, and *points*, corresponding to the divisions and subdivisions which you would have made of the head, and you will have a common measure, such as the head would have given you if it had been proportionate.'[39]

[39] Bourgelat, *loc. cit.*, p. 135.

We understand, up to a certain point, that Bourgelat may have been able to give this advice which, generally speaking, is sufficiently practical, since, in certain cases, he was able to pronounce that such a head was too small or too large. But it is always mischievous, with regard to the effect produced on the reader, to propose to him, in the application of a rule, to suppress the foundation on which this rule is established. Besides, even if all the measurements compared with the two-fifths of the length of the body are proportionate with regard to one another, the animal, in spite of this, since the head must be taken into consideration, will, in a strict sense, be none the less disproportioned.

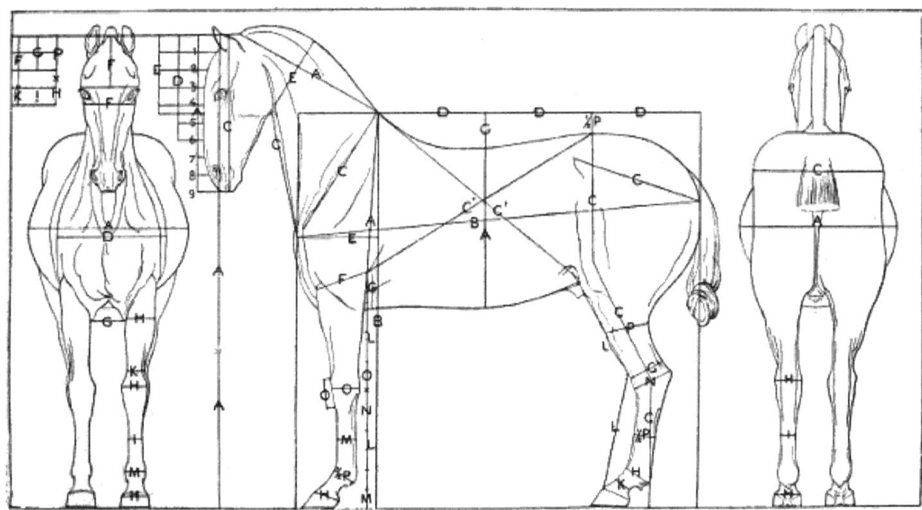

Fig. 106.—The Proportions of the Horse (after Bourgelat).

The proportions given by Bourgelat are as follows[40] (Fig. 106):

[40] *Ibid.*, p. 136, and onward.

1. **Three geometrical lengths of the head** give:

The full height of the horse, reckoned from the forelock to the ground on which he rests, provided that the head be well placed.[41]

[41] By 'the head being well placed,' Bourgelat means 'vertically posed,' the outline of the forehead then coinciding with a vertical line, which at the other end touches the anterior portion of the nose.

2. **Two heads and a half** (B)[42] equals:

The height of the body from the summit of the withers to the ground.

The length of the same body, those of the forehand and of the hind-quarter taken as a whole from the point of the arm to the point of the buttock inclusive.

[42] The letters in parentheses relate to the corresponding measures marked by the same letters on the third diagram of Fig. 106.

3. **An entire head** (A) gives:

The length of the forepart from the summit of the withers to the termination of the neck.

The height of the shoulders from the summit of the elbow to the top of the withers.

The thickness of the body from the middle of the belly to the middle of the back.

The width from one side to the other.

4. **A head measured from the top of the forelock to the commissure of the lips** (C). This measurement slightly curtailed, unless the mouth is very deeply cleft, equals:

The length of the crupper, taken from the superior point of the anterior angle of the ilium to the tuberosity of the ischium, forming the point of the buttock.

The width of the crupper or of the haunches, taken from the inferior points of the angles of the ilia.

The height of the crupper, viewed laterally, taken from the summit of the posterior angles of the ilia to the point of the patella, the leg being in a state of rest.

The lateral measure of the posterior limb, from the point of the patella, to the lateral and salient part of the ham, to the right of the articulation of the tibia with the trochlea.

The perpendicular height of the articulation above named above the ground.

The distance from the point of the arm to the angle formed by the junction of the head and neck.

The distance from the summit of the withers to the junction of the neck with the thorax.

5. **Twice this last measure** (C)[43] gives almost:

The distance of the summit of the withers to the tip of the patella.

The distance of the point of the elbow to the summit of the crupper or the posterior angles of the ilia.

[43] The proportions given in the two paragraphs 6 and 7 are, under another form, the same as those pointed out in paragraph 2, with this difference, that in this latter they are more clearly expressed.

6. **Three times this measure, plus a half-width of the pastern, the equivalent of two heads and a half**, will give:

The height of the body, taken from the top of the withers to the ground.

Its length, taken from the point of the arm to the point of the buttock inclusive.

7. **This same measure, plus the entire width of the pastern**, gives:

The total length of the body, taken accurately.

8. **Two-thirds the length of the head** (D) will equal:

The width of the chest, from the tip of one arm to that of the other, from outside to outside.

The horizontal measurement of the crupper taken between two verticals, of which one forms a tangent to the buttock, and the other passes through the summit of the crupper and touches the tip of the patella.

The third of the length of the hind-quarter and of the body taken together, as far as the vertical from the withers, touching the elbow.

The anterior length of the hind-limb, taken from the tuberosity of the tibia to the fold of the ham.

9. **One-half of the length of the head** (E) is the same as:

The horizontal distance from the tip of the arm to the vertical line from the summit of the withers and touching the elbow.

The width of the neck, viewed laterally, taken from its insertion in the trough of the jaw to the roots of the first hairs of the mane, on a line which forms with the superior contour two equal angles.

10. **One-third of the entire length of the head** (F) gives:

The height of its superior part from the summit of the forelock to a line which passes through the most salient points of the orbits.

The width of the head below the lower eyelids.

The lateral width of the forearm, taken from its anterior origin to the point of the elbow.

11. **Two-thirds of this length**[44] (G) gives:

The distance of the point of the elbow above the plane of the lower surface of the sternum.

The depression of the back in relation to the summit of the withers.

The lateral width of the posterior limbs near the hams.

The space or distance of the forearms from one ars[45] to the opposite.

[44] That is to say, two-ninths of the whole length of the head.

[45] We call the region where the superior and internal part of the forearm is joined to the trunk the 'ars.' The space between the ars of one side and the ars of the opposite side is called the 'inter-ars.'

12. **One-half of the third of the entire length of the head**[46] (H) equals:

The thickness of the forearm, viewed from the front, and taken horizontally from the ars to its external surface.

The width of the crown of the fore-feet whether from one side to the other, or from before backwards.

The width of the crown of the hind-feet, from one side to the other only.

The width of the posterior fetlocks, taken from the front to the origin of the spur.

The width of the knee seen from the front. Note: this measure is a little too large.

The thickness of the ham. Note: this measure is a little under the mark.

[46] That is to say, one-sixth of the total length of the head.

13. **One-fourth of the third of the length of the head**[47] (I) gives:

The thickness of the canon of the fore-limb: that of the hind-quarter is a little thicker.

[47] That is, one-twelfth of the length of the head.

14. **One-third of this same measure**[48] (K) equals:

The thickness of the fore-limb close to the knee in its narrowest part.

The thickness of the posterior pasterns, viewed laterally.

[48] That is, a ninth of the length of the head.

15. **The height from the elbow to the fold of the knee** (L) is the same as:

The height from this same fold to the earth.

The height from the patella to the fold of the ham.

The height from the fold of the ham to the crown.

16. **The sixth part of this measure** (M) gives:

The width of the canon of the fore-limb, viewed laterally, in the middle of its length.

The fetlock, viewed from the front.

17. **The third of this same measure** (N) is very nearly equal to:

The width of the ham, from the fold to the point.

18. **A fourth of this measure** (O) gives:

The width of the knee, measured laterally.

The length of the knee.

19. **The interval between the eyes from one great angle to the other** (P) equals:

The width of the hind-leg, viewed laterally, from the cleft of the buttocks to the inferior part of the tuberosity of the tibia.

20. **One-half of this interval between the eyes** (1⁄2 P) gives:

The width of the posterior canon-bone, viewed laterally.

The width of the fetlock of the fore-limb, from its anterior summit to the root of the spur.

Finally, the difference of the height of the crupper with respect to the summit of the withers.

It is certain that the multiplicity of these proportions, and above all the exaggeration of details into which Bourgelat fell in indicating certain of the measures which constitute the bases of some of them, may repel the reader.

For this cause we desire to add to the preceding, and also because the question which we are treating would be incomplete without it, the results obtained and published by other more modern authors, and in particular by Colonel Duhousset.[49]

[49] E. Duhousset, 'Le Cheval,' Paris, 1881.

This author, one of whose constant occupations is the measurement of the different regions of the horse, has the incontestable merit of having drawn attention to this question, and of having strained all his energies in the propagation of the knowledge which until then was little diffused. Among the proportions which he recommends, there are some which are the result of his own observations; whilst others, which he has verified and adopted, are the result of a judicious selection of those given by Bourgelat, which we have just reproduced in the preceding pages.

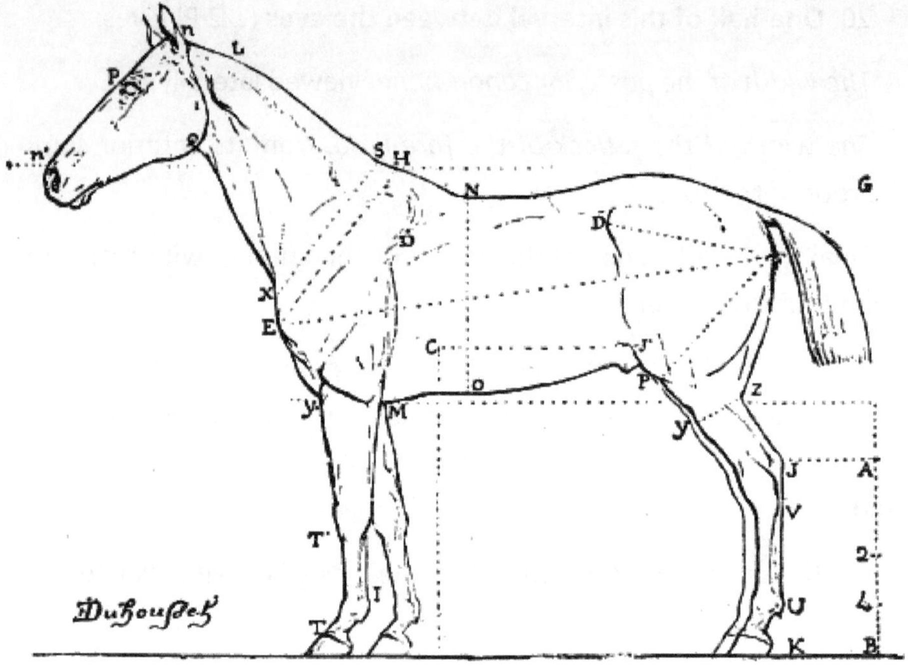

Fig. 107.—Proportions of the Horse (after Colonel Duhousset).

We join thereto also certain indications furnished by MM. A. Goubeaux and G. Barrier,[50] distinguishing these latter by the initials (G. and B.) of their authors (Fig. 107).

[50] Armand Goubeaux and Gustave Barrier, 'De l'extérieure du Cheval,' Paris, 1882.

The length of the head almost exactly equals:

1. Depth from the back to the belly, N, O,[51] the thickness of the body.[52]

[51] Look for the points indicated by these letters on Fig. 107, which is related to the proportions which are here discussed.

[52] The proportion previously indicated by Bourgelat.

2. From the summit of the withers to the point of the arm, H, E.

3. From the superior fold of the stifle to the point of the ham, J', J.

4. From the point of the ham to the ground, J, K.

5. From the dorsal angle of the scapula to the point of the haunch, D, D.

6. From the passage of the girth to the fetlock, M, I, or higher in large horses and racers; to the middle of the fetlock or lower for small ones and those of medium size.

7. From the superior fold of the stifle to the summit of the crupper in those specimens whose coxo-femoral angle is very open. This distance is always much less in others (G. and B.).[53]

[53] A proportion relative to the same region, and which at the outset might appear similar, is pointed out by Bourgelat. But there exists a difference, for Bourgelat compared the length of the head, measured from the forelock to the commissure of the lips, and not that of the entire head, to the distance which separates the summit of the rump and the tip of the patella.

Two and a half times the length of the head gives:

1. The height of the withers, H, above the ground.[54]

[54] This proportion is that given by Bourgelat.

2. The height of the summit of the crupper above the ground.[55]

[55] Consequently the withers and the crupper, being the same height, are situated on the same horizontal plane. Bourgelat, on the contrary, points out a difference of level in connection with these regions. According to him the summit of the crupper is situated below the horizontal plane passing the withers, and this distance equals half of the space which separates the great angle of one eye from that of the other.

3. Very often the length of the body, from the point of the arm to that of the buttock, although for a long time the type of Bourgelat had been set aside as a conventional model, short and massive.[56]

[56] See

And M. Duhousset adds to this:

'The drawing that we offer, which has two heads and a half in height and length, is that of a horse which we frequently meet with' (see Fig. 107; where we again consider this question of the length of the body of the horse).

'The crupper, from the point of the haunch to that of the buttock, D, F, is always less than that of the head. This difference varies from 5 to 10 centimetres. The width of the crupper, from one haunch to the other, often very slightly exceeds its length.' MM. Goubeaux and Barrier add that frequently it equals it.[57]

[57] If we refer to the proportions indicated by Bourgelat, we shall find that the proportions relative to the crupper are also indicated there.

'The crupper, such as we have just defined it, D, H, may also be found to a fair degree of exactness, as regards length, four times on the same horse.'

1. From the point of the buttock to the inferior part of the stifle, F, P.

2. The width of the neck, a little in front of the withers to a little above the point of the arm, S, X.[58]

[58] MM. Goubeaux and Barrier replace this by the following: 'The width of neck at its inferior attachment from its insertion into the chest to the origin of the withers, S, X.' Bourgelat discovered the same proportion.

3. From this latter point to below the lower jaw, X, Q, when the head is naturally placed parallel to the shoulders, E, H.[59]

[59] MM. Goubeaux and Barrier replace this by the following: 'From the insertion of the neck into the chest to the lower border of the lower jaw, X, Q, when the head is parallel to the shoulder.'

4. From the nape to the nostrils, *n, n'*.[60]

[60] MM. Goubeaux and Barrier add: 'Or to the commissure of the lips.' It is thus, besides, that Bourgelat measured the head for comparison with the crupper.

The measure of **half of the head** also acts as a good guide for the construction of the horse, when we know that it frequently applies to many of the parts—to wit:

1. From the forehead above the eyes, perpendicular to the line which is tangent to the lower jaw, P, Q.

2. Outline of the neck at the level of the base of the head, Q, L.[61]

[61] Proportion indicated by Bourgelat.

3. From the crown of the fore-foot to below the knee, T, T'.

4. In the legs, from the base of the fetlock to that of the ham, U, V.

5. Finally, it is nearly of the length of the humerus from the point E to the radius.[62]

[62] MM. Goubeaux and Barrier replace these by the following:

1. 'From the most prominent part of the lower jaw to the profile of the forehead above the eye, P, Q (thickness of the head).

2. 'From the throat to the superior border of the neck behind the nape, Q, L (attachment of the head).

3. 'From the inferior part of the knee to the crown, T, T'.

4. 'From the base of the ham to the fetlock, U, V.

5. 'Finally, from the point of the arm to the articulation of the elbow (approximate length of the arm).'

PROPORTIONS OF THE HEAD OF THE HORSE[63]

Although it is very difficult, says M. Duhousset, when we speak of measurements taken on the living animal, to formulate other than approximations, we believe we have determined with sufficient accuracy the following results, which are the outcome of our numerous observations. The head which we present is that of a horse which we have frequently come across as a mean term between the highly bred and the draught horse. Under this heading, it will not be devoid of interest to accompany with dimensions the two drawings to which are consigned the measurements in question.

[63] Extract from the work of MM. Goubeaux and Barrier on the exterior of the horse. As before, the initials G. and B. of these authors are added.

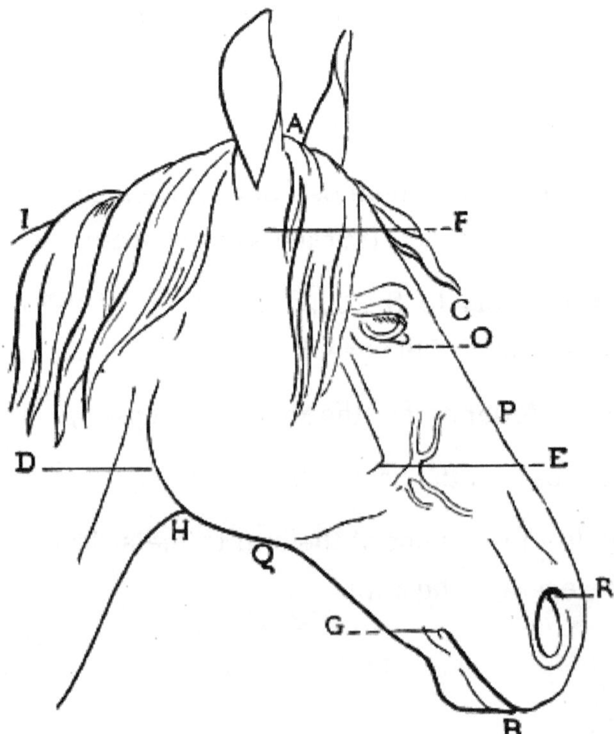

Fig. 108.—Proportions of the Head of the Horse, viewed in Profile (after Colonel Duhousset).

Head viewed in Profile (Fig. 108).—Length, A, B, from the nape to the margin of the lips, 0·60 metre.

Thickness, C, D, from the angle of the lower jaw to the anterior surface (a half-head), 0·30 metre. This line passes through the middle of the eye, taken perpendicularly, to the profile of the anterior surface. Many common horses present it, especially the heavier draught horses; in finely-bred subjects it is a little shorter (G. and B.).

Depth, I, H, of the neck in its narrowest part (a half-head), 0·30 metre. It is frequently greater; this is noticeable in all instances where the superior parts of the neck are deficient in fineness. It is this which we see in draught horses, and in those which become too fleshy (G. and B.).

Distance, O, R, of the internal commissure of the eye from the superior border of the commissure of the nostril (G. and B.) (a half-head), 0·30 metre. It is more considerable on the common head, and on that which is too long.

Distance, A, O, from the nape to the internal angle of the eye, 0·22 metre. This distance is equivalent to the thickness of the head, P, Q, taken perpendicularly from the profile of the anterior surface, and passing at the level of the maxillary fissure and spine.

It is, again, equal to Q, O, from the internal angle of the eye to the maxillary fissure; and to P, G, from the middle of the face to the commissure of the lips (G. and B.).

The distance, P, E, from the middle of the face to the maxillary spine is about the sixth of the total length of the head—0·10 metre.

The line B, E, reckoned from the extremity of the lips to the maxillary spine, is equal:

To E, F, from the maxillary spine to the external auditory meatus, to be seen only on the skull;

To H, G, from the insertion of the neck in the trough to the commissure of the lips (G. and B.);

To Q, R, from the maxillary fissure to the superior commissure of the nostril (G. and B.);

To Q, B, from the fissure of the maxilla to the border of the lips (G. and B.);

To O, D, from the internal angle of the eye to the angle of the lower jaw, provided that the line C, D be in proportion (G. and B.).

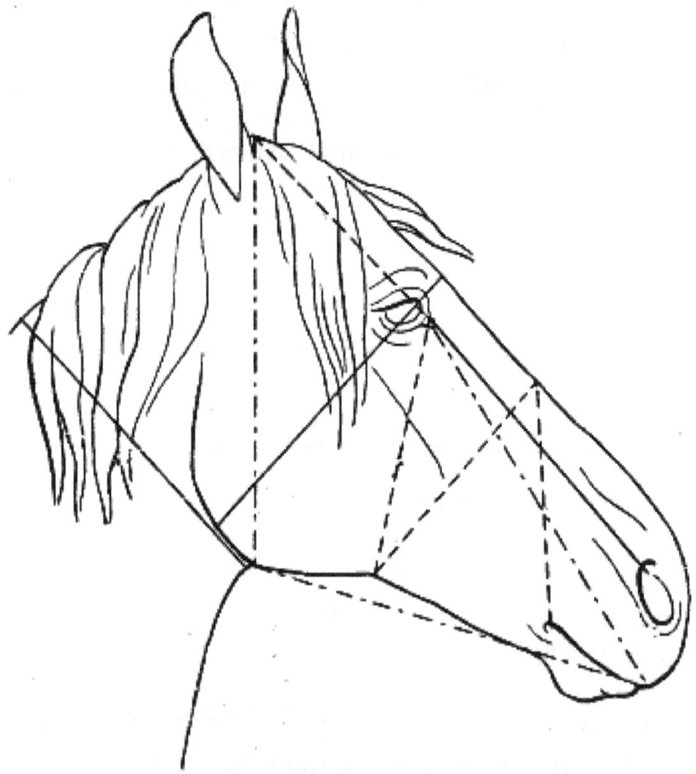

Fig. 109.—The Same Design as that of Fig. 108, on which we have indicated, by Similar Lines, the Principal Corresponding Measurements.

Half the length of the head, and the dimensions which equal it; distance which separate the nape from the internal angle of the eye, and the dimensions which equal it; distance which separates the internal angle of the eye from the border of the lips, and the dimensions which equal it.[64]

[64] It is thus that in our teaching, but by means of lines of different colours, we present the proportions reproduced in Fig. 108. Experience has demonstrated to us that this replacement of letters by conventional lines renders the proportions more easily appreciable, and that these lines, striking the eye more forcibly, then impress themselves better on the memory. Fig. 111 bears the same relation to Fig. 110.

Finally, very frequently to O, H, from the internal angle of the eye to the insertion of the throat into the maxillary trough (G. and B.).

An equality still more frequent is that which exists between the distances:

O, B, from the internal angle of the eye to the margin of the lips;

A, H, from the nape to the insertion of the throat into the maxillary trough;

And H, B, from this latter point to the margins of the lips.

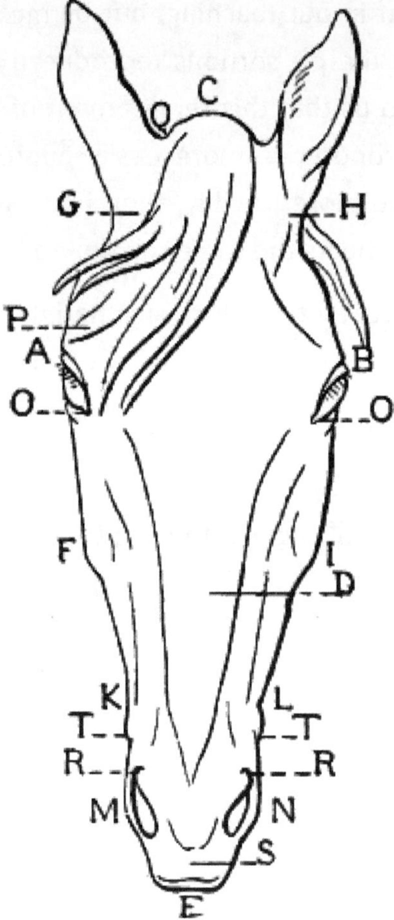

Fig. 110.—Proportions of the Head of the Horse, seen from the Front (after Colonel Duhousset).

The Head, Front View (Fig. 110).—If, to continue our examination, adds M. Duhousset, we regard the head from the front, we find its greatest width at A, B, the extreme points of the orbital arches.

This width is 22 centimetres.

It is again equal to:

A, C, from one arch to the nape;

A, D, from one arch to the middle of the face.

D, E, from the middle of the face to the margin of the lips.

From the auditory canal, G, to the maxillary spine, F, is the same distance as from this point to the margins of the lips, E, or, better, to the end of the teeth.

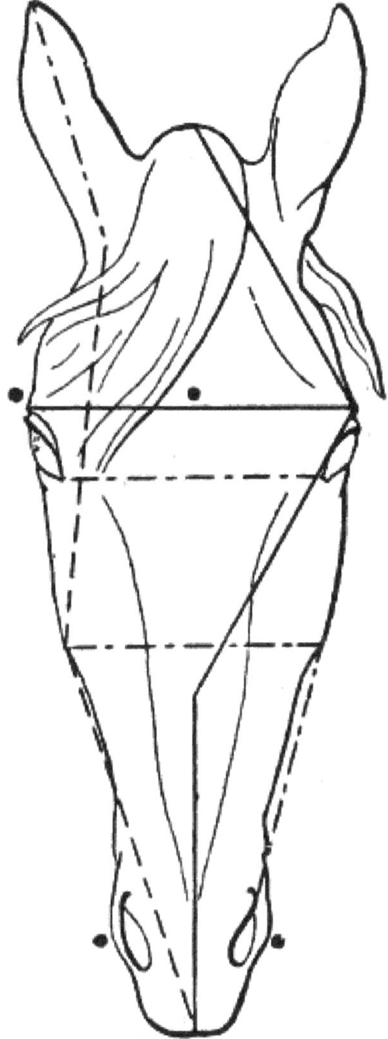

Fig. 111.—The Same Figure as Fig. 110, on which we have marked by Similar Lines the Principal Measurements which correspond thereto.

Distance which separates one of the orbital arches from that of the opposite side, and the dimensions which equal it; distance which separates the auditory meatus from the maxillary spine, and the dimensions which equal it; distance which separates one maxillary spine from that of the opposite side, and the dimensions which equal it; distance which separates the lip of one side from that of the opposite, and the dimensions which equal it.[65]

[65] See the note relative to Fig. 109.

The line G, C, from the auditory meatus to the nape, is equal to the sixth of the head, 10 centimetres; the line A, G, from the orbital arch to the auditory meatus, is a little longer, and measures 12 centimetres.

The distance F, I, comprised between the maxillary spines, is 18 centimetres.

It is equal to:

O, O, the distance between the internal angles of the eyes (G. and B.);

F, R, the distance from the maxillary spine to the superior commissure of the corresponding nostril (G. and. B.);

F, P, from the maxillary spine to the *salt-cellar*.[66]

[66] We designate under the name *salt-cellar* a depression situated external to the frontal region and above the eye.

From the nape to the internal angle of the eye, C, O, is the same distance as from this latter point to the commissure of the lips, O, T; and from the maxillary spine to the upper lip F, S (G. and B.).

The distance apart, T, T, of the two commissures of the lips gives, very nearly, the distance from the superior border of the orbital arch to the base of the ear or the auditory meatus. In the state of rest, the outer limit of the separation of the nostrils does not exceed the width of the knee;[67] we frequently find the same distance intercepted above the nape by the tranquil ears. In the figure (Fig. 110) we have intentionally represented them directed in a different plane, in order to show that when the pinna is turned backward, it none the less preserves the contour of bracket form, more or less pronounced according to the breeding of the subject, and characterizing in repose the interior curves of the ear.

[67] We remind our readers that the name 'knee' is given by veterinarians to the region occupied by the carpus.

The extreme limit of the lips, M, N, but very slightly exceeds that of the nostrils; on many heads of harmonious proportions this distance is found to be the half of A, B.

In order not to interrupt the course of the preceding exposition, we decided to withhold till afterwards some reflections which have been suggested to us by certain of the proportions which are there indicated. The proportions in question are important—we may even say that they are fundamental, for they have for object the relation which exists between the length of the head, the height of the body, and the length of the latter.

We have already seen that, according to Bourgelat, the length of the head is contained two and a half times in the length of the body, from the point of the arm to the point of the buttock; and, also, two and a half times in the height measured from the apex of the withers to the ground. We saw afterwards that M. Duhousset, having adopted these proportions, pointed out, further, that the same dimension was again found equally to exist from the summit of the crupper to the ground—a height which Bourgelat considered as being of less extent. There results, then, from the latter proportions, which we have just recalled, this interesting fact: that they simplify very much, from the point of view of design, the placing in position of the horse, on the condition always that this latter be always viewed directly on one of its lateral aspects.

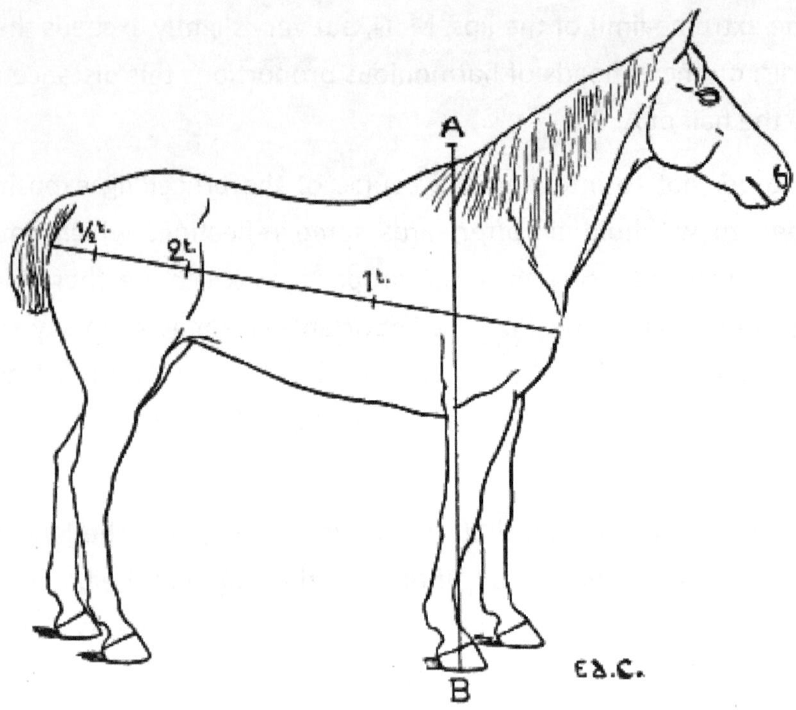

Fig. 112.—Horse of which the Length contains more than Two and a Half Times that of the Head, and of which this Dimension (A, B) exceeds the Height.

Indeed, in this case, if we except the neck and the head, the body, inasmuch as its height and its length are equal, may be inscribed in a square, of which one of the sides corresponds to the withers and to the summit of the crupper, two of the other sides to the point of the arm and to that of the buttock the fourth being represented by the ground. This is simple, but this simplicity even has its inconveniences.

It follows that this proportion, thus expressed, seems to exclude from every artistic representation certain categories of horses, which upon the whole might be regarded as beautiful, and the existence of which in any case it would be a pity not to indicate.

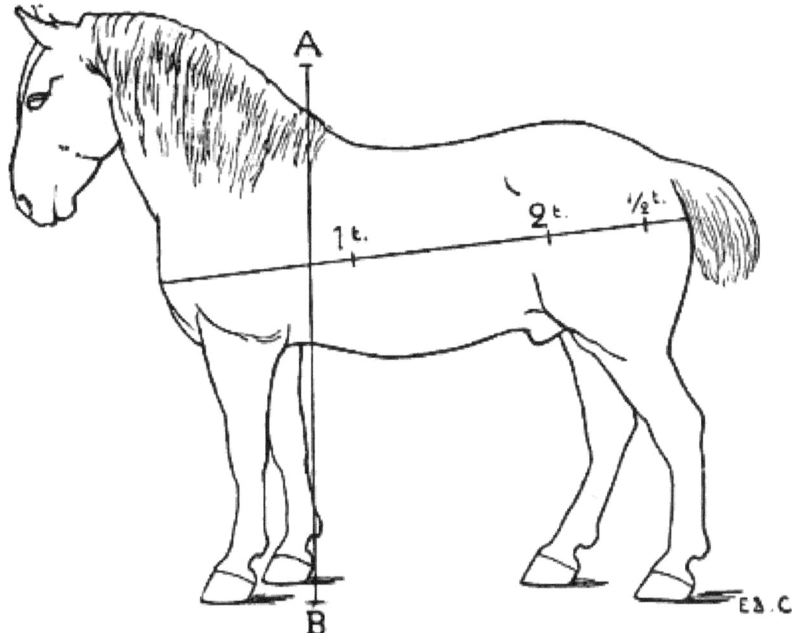

Fig. 113.—Horse of which the Length contains more than Two and a Half Times that of the Head, and of which this Dimension (A, B) exceeds the Height.

Let us examine at the outset that which is relative to the length of the body, equal to two and a half times the length of the head. This proportion is sometimes met with, and therefore may be considered exact; but it is necessary to add that its existence is not discoverable in the majority of cases. That for some authors it constitutes a perfect model we will not gainsay, but it is our impression that, when it exists, the head appears a little large, or, more exactly, the body a little short.

Without attaining exactly to three times the length of the head, as some authors (Saint-Bel, Vallon) have announced, the body of the horse, nevertheless, measured as is stated above, frequently contains it more than two and a half times. We give in support of this some outline reproductions, executed after photographs (Figs. 112, 113, 114).

There still remains the question regarding the equality of the height and of the length of the body of the horse.

This equality, after the proportions previously indicated, would seem bound to appear in all the cases observed. Now, if we measure the examples reproduced in Figs. 112, 113, and 114, we shall see that sometimes the two dimensions are unequal, the height being greater than the length, or inversely.

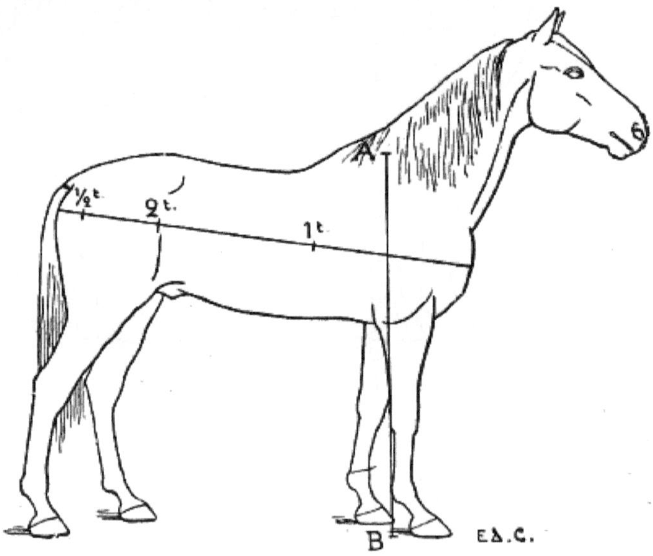

Fig. 114.—Horse of which the Length contains more than Two and a Half Times that of the Head, and of which this Dimension (A, B) is Inferior to the Height.

It is the same, if we examine a certain number of specimens; we are able to determine that the proportion chosen in preference by authors is not exactly that which is oftenest met with. It will, very probably, be objected that it is so for the most beautiful types, and that the indifferent ones are generally the more numerous. The essential thing would be to know, above all, if the type of two heads and a half of length and of height is really the only beautiful one. However that may be, of the fifty African horses measured by M. Duhousset, only fourteen possessed the equality indicated; twenty-six were less long than high, and ten more long than high.[68]

[68] E. Duhousset, 'The Horse,' Paris, 1881.

CHAPTER V

THE PACES OF THE HORSE

As a completion of the studies we have just been making, some notions relative to the paces of the horse seem to us to be absolutely indicated.

Let it be permitted to us to remind the reader in this connection that we have already been for twenty-one years occupied with this question, and that by means of an articulated figure, a sort of movable mannikin, we have endeavoured to demonstrate to artists the differences which characterize the various paces of the horse.[69] The arrangement then employed cannot, evidently, be used in the present volume, but we will inspire ourselves, in the preparation of the present chapter, with the elements of demonstration which we have employed, and which, in the course of our teaching, we have had the satisfaction of seeing favourably received.

[69] Édouard Cuyer, 'Les Allures du Cheval,' demonstrated with the aid of a coloured, separable, and articulated table, Paris, 1883.

This table was the subject of a note communicated to the Academy of Sciences by Professor Marey ('Comptes rendus de l'Académie de Sciences') at the meeting of June 26, 1882. On the other hand, it has been the subject of a presentation which we have had the honour of being permitted to make to the Academy of Fine Arts at the meeting of November 4, 1882.

The fasciculus in question has been since united with a more complete whole as regards the study of the horse. E. Cuyer and E. Alex, 'Le Cheval:

Extérieur, Structure et Fonctions, Races,' avec 26 planches coloriées, découpées et superposées, Paris, 1886.

The progressive movements by which an individual transports himself from one place to another do not operate according to a unique method and with a constantly uniform velocity. These various modes of progression are designated under the name of *paces*.

It is extremely difficult to analyze, by simple observation, the movements which characterize these gaits. Let us, for example, examine the displacements made by the limbs of a horse during that of walking; if we have no notion of these displacements, it will be, so to speak, impossible to determine in what order they are executed. The sight of the imprints left on the ground by the hoofs is not a sufficient means of demonstration, especially for artists. The noise made by the blows of these limbs, or by the little bells of different timbre suspended from them, are absolutely in the same case.

Processes enabling us to fix or to register the paces are in every way preferable. Such really exist; they are: instantaneous photography and those which constitute the graphic method of Professor Marey. The results given by the photograph are certainly appreciable; but, from the didactic point of view, we give the preference to the graphic method, the general characters and the mode of application of which we now proceed to analyze.[70]

[70] We cannot too strongly recommend the reading of the excellent works which Professor Marey has published, and which have for their object the study of movements, as well as the exhibition of the procedures which he has employed. E. J. Marey, 'La Machine Animale,' Paris, 1873; 'La Méthode graphique dans les Sciences expérimentales,' Paris, 1884; 'Le Vol des Oiseaux,' Paris, 1890; 'Le Mouvement,' Paris, 1894.

It is necessary to understand first of all, in this connection, that which relates to a man's walking pace.

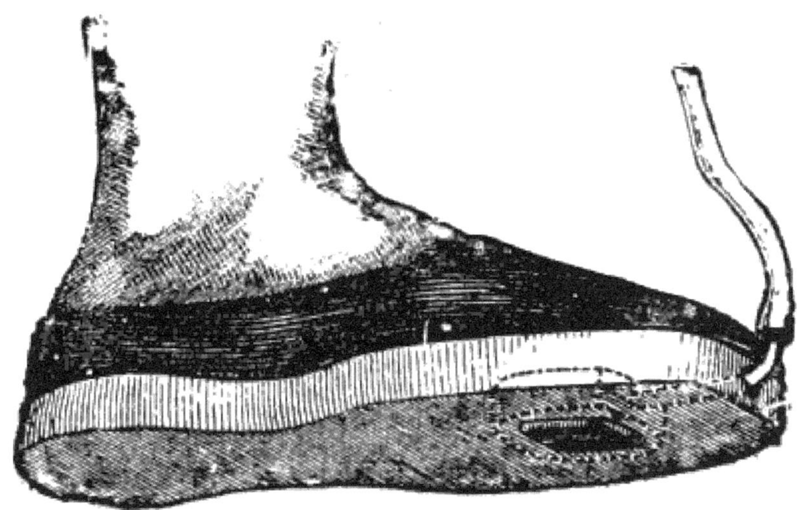

Fig. 115.—Experimental Shoes, intended to Record the Pressure of the Foot on the Ground.

The method of Professor Marey rests on the following principle: Suppose two rubber globes connected with one another by a tube. If we compress one of these globes, the air which it contains will be driven into the other, and will afterwards return when the pressure has ceased. Nothing more simple, evidently; but it is necessary to describe it in detail in order the better to comprehend that which follows: The walker who is the subject of experiment is furnished with special shoes (Fig. 115), having thick indiarubber soles, hollowed in the interior, so that the whole thus constituted forms a sort of hollow cushion which is compressed under the influence of the pressure of the foot on the ground. A tube which is attached to a registering apparatus, which the person who is walking carries in his hand, communicates with this cavity (Fig. 116). This apparatus is formed of a metal drum, which is closed at its upper part by a flexible membrane. Each time that one of the man's feet presses on the ground, the air contained in the cavity of the sole of the shoe is driven into the drum, which we have just mentioned, and the flexible membrane of this drum is elevated. To this membrane is attached a vertical rod which supports a horizontal style.

Fig. 116.—Runner furnished with the Exploratory and Registering Apparatus of the Various Paces.

When the membrane, as we have just seen, is elevated, the style is lifted, and then descends when the pressure of the foot ceases. It traces these displacements on a leaf of paper, the surface of which is covered with a thin layer of lamp-black, which it removes by its contact; different parts of this surface are successively presented to it, the paper being rolled round a cylinder which is turned on its axis by means of a clockwork movement. It is necessary to add that the inscription is made, in the study of the walk of man, by means of two styles, each corresponding to one of the feet.

The tracings thus obtained, which are read from left to right, are sufficiently simple; but to understand them properly, it is necessary to remember that the style undergoes a movement of ascensional displacement during each pressure of a foot, and that, on the other hand, it descends when the latter is separated from the ground. We also see, on the tracing which it leaves, a line which ascends and then descends; the meaning of this is that first the foot presses on the ground, and is afterwards raised from it.

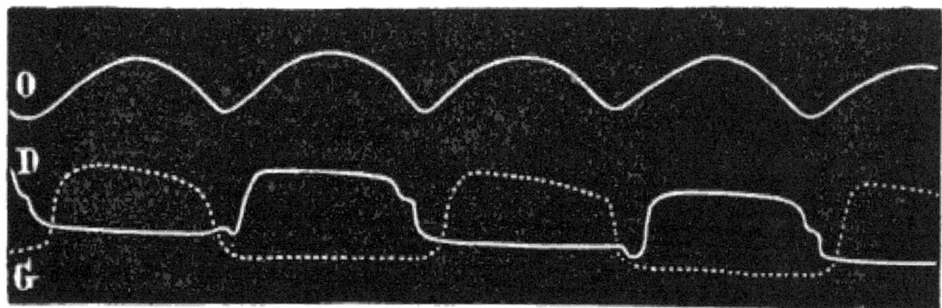

Fig. 117.—Tracing of the Running of a Man (after Professor Marey.)

D, Pressures and elevations of the right foot; G, pressures and elevations of the left foot.

On the tracing (Fig. 117), the line D relates to the right foot; the line G, which is dotted so that it may not be confused with the preceding, corresponds to the left foot. The line G first ascends; the meaning of which is that the left foot presses on the ground; afterwards it descends: this indicates that the pressure of the foot has ceased. It is the same for the right foot. As we see, the pressures succeed each other; when the left foot touches the ground, the right is separated from it; when the latter presses the ground, it is the left which no longer rests there.

The line O is related to the movements of the body, as indicated by the oscillations of the head. We will neglect these.

But this tracing, which serves us for an example, is not, it must indeed be said, of very easy reading; it would be still less so if the paces of a horse were registered, for there would then be four lines, the entanglement of which would cause greater complication.

These difficulties of reading need be no longer feared, if we transform the tracing into a notation by means of the following diagram.

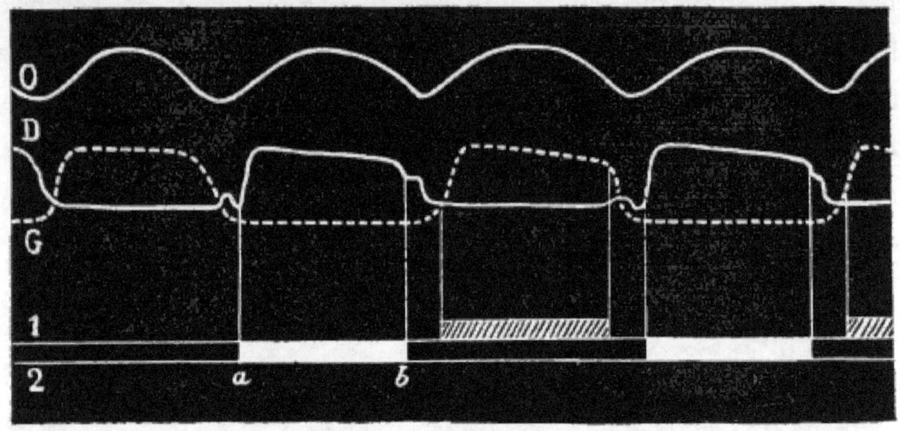

Fig. 118.

There are drawn (Fig. 118) below the graphic tracing two horizontal lines (1, 2). From the point where the line D rises (commencement of the pressure of the right foot), and from the point where this same line descends (end of the same pressure), we let fall two vertical lines joining the two horizontal ones mentioned above. At this plane, and between the two vertical lines, we mark a broad white one (*a, b*). This expresses, by its length, the duration of the period of pressure of the right foot. In doing the same for the line G, we obtain for the indication of a pressure of the left foot an interval of the same kind, in which are marked cross-lines, or which is tinted gray, in order to avoid all confusion with the preceding tracing.

This notation can, with sufficient exactitude, be compared to that which is employed in the musical scale. The horizontal lines 1 and 2 represent the *compass*. We there also see *notes*; these are the bars indicating the pressure, of which the value—that is to say, the duration—is represented by the length of these bars. It is the same with regard to the intervals of *silence*: these are expressed by the intervals which separate the pressures, and correspond to the moments in which, during certain paces, such as running, the body is raised from the ground.

Besides, we see intervals of this kind on the notation reproduced (Fig. 118) relative to the running of man.

Fig. 119.

In order to make the signification of these tracings still better understood, we reproduce four varieties of them (Fig. 119).

The first notation is that of ordinary walking. The pressures succeed each other regularly.

The second shows what takes place during the ascent of a staircase. At a certain moment, the weight of the body is upon both feet at the same time, one of them not quitting the lower step, until the other is already in contact with the step above. Accordingly, there is thus produced an overriding of the pressures.

The third is relative to running, and has already been represented in Fig. 118. The pressures of the feet are separated by the times of suspension.

The fourth also represents running, but in this case more rapid and characterized by the shorter pressures, the slightly longer periods of suspension intervals, and the quicker succession of movements.

Before putting aside the indications relating to the walking movements of man—indications which it was necessary to give in order to render intelligible those which are connected with the paces of the horse—we have yet to fix the value of that which we call 'a step.'

It is generally admitted that a step is constituted by the series of movements which are produced between the corresponding phases of the action of one foot and that of the other—for example, between the moment at which the right foot commences its pressure on the ground and that at which the left foot commences its own. It is necessary to adopt here another method of looking at it, and to regard the preceding as being but a *half-step*. The step should then be defined as being constituted by the series of movements which are executed between two similar positions of the same foot—as, for example, between the commencement of a pressure of the right foot and the similar phase of the following pressure of the same foot. We shall soon understand the importance of this definition.

Before entering on the details of the paces of the horse, it is necessary to see how the limbs of the latter oscillate during the period of a complete step; or, which is the same thing, to determine what the displacements are which a limb executes between two similar positions of its foot.

If we examine one of the limbs during a forward movement of the animal, we see that this limb passes through two principal phases: (1) It is raised from the ground; (2) it resumes contact with the ground. Each of these phases is divided into three periods of time, which we proceed to analyze in connection with the anterior limb.

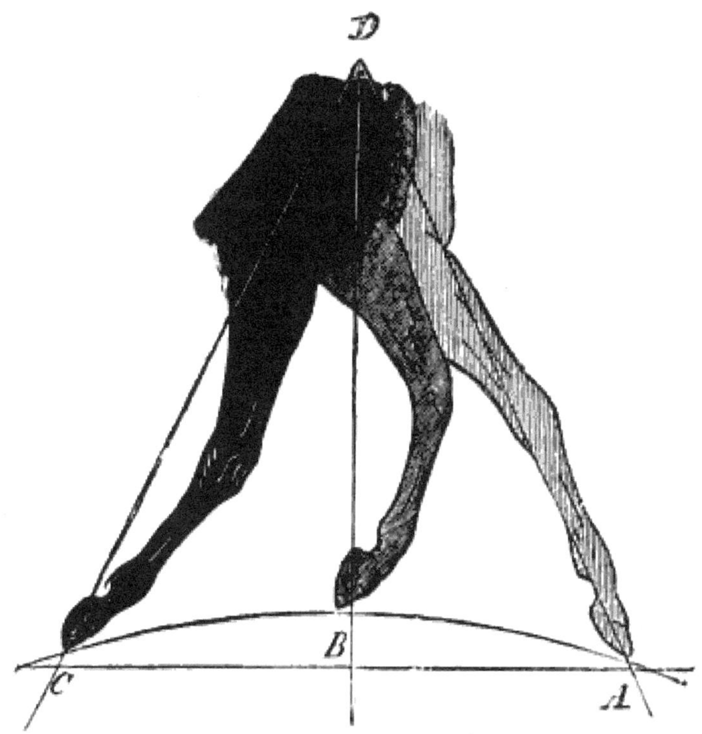

Fig. 120.—Swing of the Raised Anterior Limb (after G. Colin).[71]

C, Lifting; B, suspension; A, placing.

[71] G. Colin, 'Traité de Physiologie Comparée des Animaux,' third edition, Paris, 1886.

The foot quits the ground (Fig. 120, C); this may be called *lifting*; the limb is oblique in direction downwards and backwards. This same limb is flexed and carried forward (Fig. 120, B), and, as it is supported by the action of its flexors, this is the period named *suspension*; the hoof is vertical. Then the limb is carried still further forward, becoming extended (Fig. 120, A); the heel is lowered, and the foot, being oblique, is directed towards the ground; this is the *placing*.

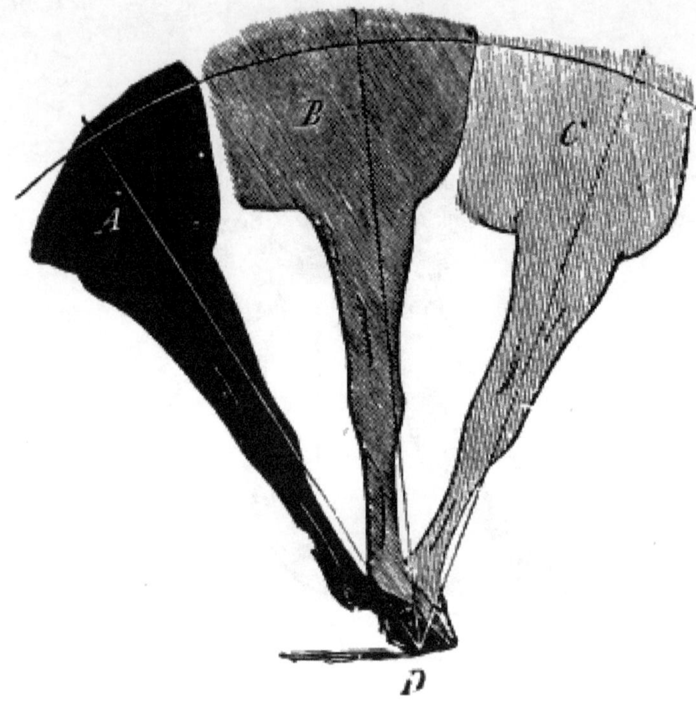

Fig. 121.—Swing of the Anterior Limb on the Point of Pressure (after G. Colin).

A, Commencement of the pressure; B, centre of the pressure; C, termination of the pressure.

Then takes place pressure (Fig. 121). The foot has just been placed on the ground; the limb is oblique in direction downwards and forwards; this we call *commencement of the pressure* (Fig. 121, A). Then the body, being carried forward, whilst the hoof, D, is fixed on the ground, the limb becomes vertical: this stage is *mid-pressure* (Fig. 121, B). Finally, the progression of the body continuing, the limb becomes oblique downwards and backwards; it is now at the *termination of pressure* (Fig. 121, C), and proceeds to lift itself anew if another step is to be made.

In conclusion, the inferior extremity of the limb describes, from its elevation to its being placed on the ground, an arc of a circle around its superior extremity (Fig. 121, D); whilst, during the pressure, it is its superior extremity which describes one around its inferior extremity, then fixed on the ground (Fig. 121, D).

If we simultaneously examine the two fore-limbs, we remark that when one of them begins its pressure the other ends it, and *vice versâ*.

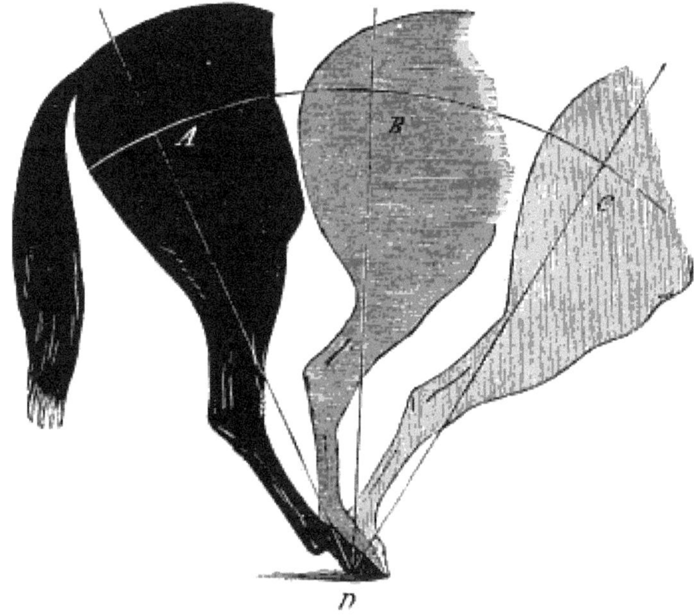

Fig. 122.—Posterior Limb, giving the Impulse (after G. Colin).

A, Commencement of pressure; B, centre of pressure; C, termination of pressure.

As to the hind-limbs, the oscillations are similar to those of the fore ones. In the second half of the pressure—that is, when they are passing from the vertical direction (Fig. 122, A) to extreme obliquity backwards (Fig. 122, C)—the effect of their action is to give propulsion to the body.

The fore and hind limbs make the same number of steps, and the steps have the same length.

The limbs of any quadruped—but we make special allusion to those of the horse—are divided into groups in the following manner:

The anterior pair constitutes the *anterior biped*. The *posterior biped* is that formed by the posterior limbs.

The name of *lateral biped* serves to designate the whole formed by the two limbs of the same side. The right fore-limb and the right hind-limb form the *right lateral biped*. The two others form the *left lateral biped*.

A fore-limb and hind-limb belonging to the opposite side form a *diagonal biped*, which also takes the name of the fore-limb which forms a part of it. Thus, *the right diagonal biped* is formed by the association of the right fore-limb and the left hind one. The *left diagonal biped* is, consequently, the inverse.

It is necessary to remember well these preliminary indications; it is the only means of comprehending with facility that which is about to follow.

Let us first return to the grouping of the limbs. The denominations *anterior* and *posterior bipeds* render clearly perceptible the comparison which consists in regarding a horse when walking as capable of being represented by two men marching one behind the other, and making the same number of steps. According as they move the legs of the same side at the same time in 'covering the step,' or march in contretemps step, we find reproduced all the rhythms which characterize the different paces of the horse.

Fig. 123.—Notation of the Ambling Gait in the Horse (after Professor Marey).

Professor Marey has studied these paces by a similar method to that which he adopted for the walking of man, and which we have already described. He employed hollow balls fixed under the hoofs, and a registering apparatus with four styles, each corresponding to one of the limbs. The tracing obtained is rather complicated, since two sets of lines are found marked. But a notation similar to that of which we have spoken can be discovered, and its exact signification should now be determined. For this purpose, we have selected the most simple (see Fig. 123). We there see, placed in two superimposed lines, the pressure markings of the right feet (white bands), and of the left feet (gray bands). On the upper line are found those related to the fore-legs; the lower lines contain those associated with the hind-legs. It is, in brief, the superposition of two notations of the human walking movements. And seeing that, as we have previously pointed out, we may make a comparison between a quadruped and two men placed one behind the other, it is easy to understand the significance of the superimposed notations, if we accustom ourselves to look on them as the notations of two bipeds.

To read these notations—that is, to learn to know what occurs at each of the movements of the pace—it is necessary, indeed, to remember that they should be examined in vertical sections; it is to each of these sections—of these vertical divisions—that each of the movements which we more particularly wish to analyze corresponds.

We proceed to study first the pace of ambling, because it is the most simple; we shall then consider the trot, and, finally, we shall examine that which is the most complicated, viz., the step.

The Amble.—To give an exact idea of the general character of the amble, let us fancy the two men whom we discussed above marching one behind the other and walking in step—that is, moving the legs of the same side simultaneously. They will thus represent the amble, which, indeed, results from the alternate displacements of the lateral bipeds; the limbs of the same side (right or left) execute the same movements in the same time.

This is what the notation indicates (Fig. 123). We there see that the pressures of the right fore-foot, marked by the white bands in the upper range, are exactly superposed on those of the right hind one, which are marked by a similar band on the lower line; this means that the pressures took place in the same time. We there see also a similar arrangement of the gray bands, which has a similar significance for the left fore and hind feet.

Fig. 124.—The Amble: Right Lateral Pressure.[72]

[72] The figures which, in the present study, reproduce the different paces, have been made from our articulated horse.

And if we recollect the three phases of pressure (Figs. 121, 122), we shall comprehend, in looking at the diagrams, that, at the initial stage (A), the limbs are commencing their pressure, and are oblique downwards and forwards; that afterwards (B) the two limbs are vertical, since they are at the middle of the pressure stage; and that finally (C) they are oblique downwards and backwards, for it is then the termination of their pressure (Fig. 124).

During the time that the right limbs are pressing (notation, white bands) the left limbs are raised; afterwards these latter take up the pressure (gray bands), and then the right limbs are raised in their turn.

During the pace of ambling the weight of the body, which is wholly sustained by the limbs of one side only, is not in equilibrium, so that the limbs which are raised return by a brisk movement to the position of support in order to re-establish it.

The Trot.—We have just seen that, in order to represent the amble, the two marchers moved their right limbs simultaneously, and then their left ones.

Let us suppose now that the hinder man anticipated by half a pace the movement of the front one, then will be found realized the association and the nature of the displacements of the limbs during the pace of the trot.

By this anticipation of a half-step (we have defined, what is to be understood by the word *step*), it follows that when the marcher who is in front advances his right leg it is the left leg of the marcher who follows him that is carried in the same direction. We should thus conclude from this that the trot is characterized by a succession of displacements of the diagonal bipeds.

Fig. 125.—Notation of the Gait of the Trot in the Horse (after Professor Marey).

Fig. 126.—The Trot; Right Diagonal Pressure.

Indeed, if we examine the notation of this gait (Fig. 125), we see that with the pressure of the right fore-foot is found associated the pressure of the left hind-foot. It is, accordingly, a typical diagonal biped (Fig. 126).

Fig. 127.—The Trot; Time of Suspension.

But it is necessary to add that these groups of pressures do not succeed one another without interruption, except in the slow trot. In the ordinary trot, or in that in which the animal's strides are very long, the body between each of the double pressures which we have just been considering is projected forward with such force that it remains for an instant separated from the ground. This is what we designate by the name of *time of suspension* (Fig. 127). The notation in this case would be slightly different from that which we reproduce above, in this sense: that between the diagonal pressures there then would be found an interval, since during the time the body is suspended none of the feet can produce a pressure-mark (see, with regard to these intervals, the notations of the running of a man, Fig. 118, and Fig. 119, 3, 4).

The Walk.—Although slow, a feature which would seem to make it possible to permit its analysis in a horse when walking, this pace is difficult to comprehend without sufficient preliminary study.

We saw above that in order to represent the amble the marchers had to move the legs of the same side simultaneously. We have also just seen that in order to represent the trot the marcher at the back had to anticipate by a half-step. Suppose, now, that this same marcher anticipates the man in front by a quarter-step only, or by a half-pressure period, and thus will be found realized the order of succession of the limbs in the gait or pace called the *walk*. The feet meet the ground one after the other, since they are each in advance by half the duration of a pressure. The strokes are four in number during the period of a step of this pace; in the amble and in the trot they do not exceed two, for then the limbs strike the ground in lateral diagonal pairs.

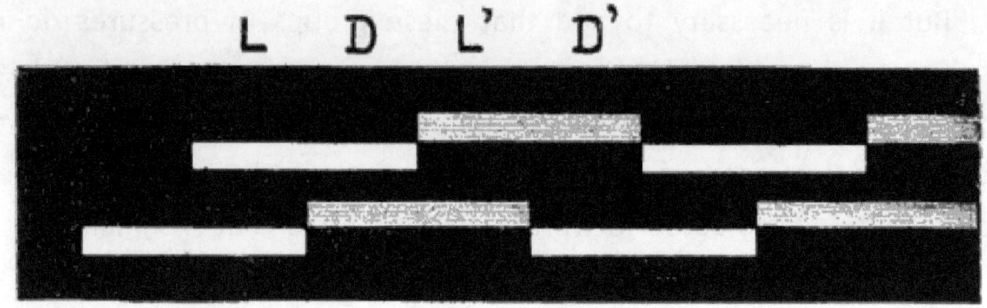

Fig. 128.—Notation of the Pace of Stepping in the Horse (after Professor Marey).

L, Right lateral pressure; D, right diagonal pressure; L′, left lateral pressure; D′, left diagonal pressure.

If we examine the notation of the pace of walking (Fig. 128), we see that the right fore-foot commences its pressure when the right hind-foot is in the middle of its own, and that the hinder left begins in the middle of that of the right fore-foot, and that it is itself at the midst of its pressure when the left fore-foot touches the ground, etc. In a word, the foot-fallings occur in the following order and at regular intervals—the fore right foot is here considered as acting first: right fore, left hind, left fore, right hind, and so on in succession.

As to the nature of the bipeds which succeed one another, it is easy to understand them by means of the notation. In reading this from left to right, we see that the associations of pressure are first made by the two right feet, then by a right foot and a left one, then by two left feet, and, finally, by a left and right. It is, accordingly, a succession this time of lateral and diagonal pressures.

Fig. 129.—The Step: Right Lateral Pressure.

Fig. 130.—The Step: Right Diagonal Pressure.

Thus, we find at the start a right lateral pressure (Fig. 129), next a right diagonal (Fig. 130), then a left lateral; finally, a left diagonal pressure. It is thus that the initial letters L, D, L′, D′ further indicate the notations represented in Fig. 128.

Fig. 131.—The Gallop: First Period.

Fig. 132.—The Gallop: Second Period.

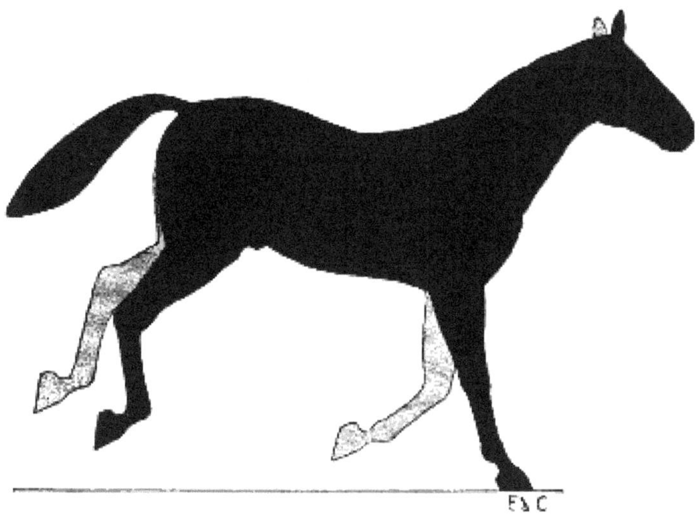

Fig. 133.—The Gallop: Third Period.

Fig. 134.—The Gallop: Time of Suspension.

The Gallop.—The ordinary gallop is a pace of three phases. The first is characterized by the fact that one hind-limb alone rests on the ground (Fig. 131); in the second the animal is on a diagonal support (Fig. 132); in the third it comes down on a fore-limb (Fig. 133). The body is then raised (Fig. 134), and to this period of suspension succeed anew the three modes of pressure indicated above.

The gallop is said to be from either right or left. In the gallop from the right, the right fore-leg is the more frequently in advance of its neighbour; it is the last to be placed on the ground. The left foot of the posterior biped is the one which commences the action.

An entirely opposite arrangement characterizes the gallop from the left.

Fig. 135.—Notation of the Gallop divided into Three Periods of Time (after Professor Marey).

1, First period; 2, second period; 3, third period.

The notation reproduced in Fig. 135 corresponds to the gallop from the right. It is there seen, as we pointed out above, that in the first phase the exclusive support of the left hind-foot takes place (1); that afterwards, in the second, commence simultaneously, the pressures of the left fore and the right hind foot (2); this is the left diagonal support; and that finally, in the third, the body comes down on a fore-limb, which is then the right (3); and that for a moment it is on this limb alone that the animal rests.

To these three phases on the notation succeeds an interval; this is the period of suspension.

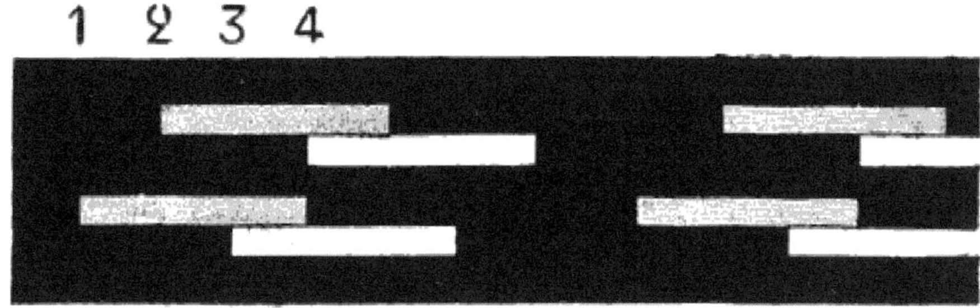

Fig. 136.—Notation of the Gallop of Four Periods in the Horse (after Professor Marey).

1, First period; 2, second period; 3, third period; 4, fourth period.

The gallop of four phases only differs from the preceding in that the foot-fallings of each diagonal biped occur at slight intervals, and give distinct sounds. The notation is reproduced in Fig. 136.

The Leap.—The leap is an act by which the body is wholly raised from the ground and projected upwards and forwards to a greater or less distance.

It is prepared for by the flexing of the hind-limbs, which, by being suddenly extended, project the body, and thus enable it to pass over an obstacle.

Fig. 137.—Leap of the Hare (after G. Colin).

This preparatory arrangement is very remarkable in the leap of the lion, the cat, and the panther, which execute springs of great length; in the horse, in which the leap is not an habitual mode of progression, this flexion of the hinder limbs is less marked. With this animal the leap is generally associated with the gallop; nevertheless, it is sometimes made from a stationary position. In observing the hare or the rabbit, in which the leap is habitual, we notice (Fig. 137) that the hind-limbs, being extremely flexed, rest on the ground as far as the calcaneum, are then straightened by the action of their extensors, become vertical and then oblique backwards at the moment the body is thrown forward into space by the sudden extension of these limbs.

The action of the extensors is energetic and instantaneous, and their energy is greater than in ordinary progression, for it is required to lift the body and to project it forcibly a more or less considerable distance. It is the extreme rapidity of this action which enables the animal to clear an obstacle, for without this condition the body would be raised, but not separated from the ground.

First of all, in reaching the obstacle to be cleared, the horse prepares to leap by taking the attitude of rearing; the hind-limbs are flexed and carried under the body, the fore-quarters are raised, and the different segments of the fore-limbs are flexed (Fig. 138).

Fig. 138.—The Leap.

Fig. 139.—The Leap.

Fig. 140.—The Leap.

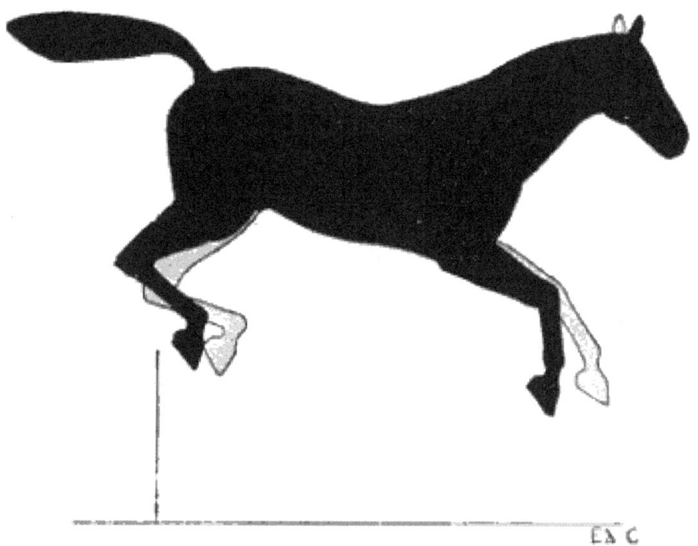

Fig. 141.—The Leap.

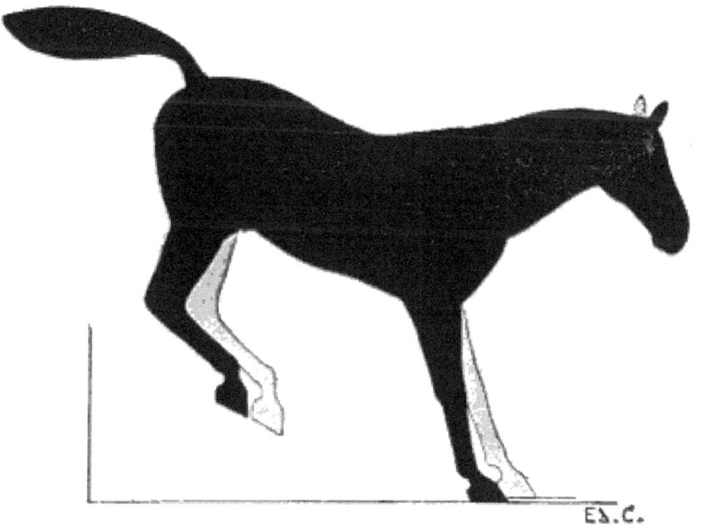

Fig. 142.—The Leap.

Fig. 143.—The Leap.

One sudden trigger action produced by the violent contraction of the extensors of the hind-legs then takes place, and the animal is projected forwards, while he flexes the fore-legs more and more (Fig. 139). He has then risen above the obstacle (Fig. 140). Then while he makes the downward and forward balancing movement, and points his fore-limbs in the same direction, he flexes the hind ones (Fig. 141). Whilst the latter are further flexed, in order to pass the obstacle in their turn, the fore-limbs which are extended come into contact with the ground (Fig. 142). Finally, in the last phase of the leap, the animal, raising himself in front, after the impact of his hind-feet has taken place (Fig. 143), prepares to continue the pace at which he progressed before meeting the obstacle which he had to clear.

London: Baillière, Tindall and Cox, 8,

Henrietta Street, Covent Garden, W.C.

Ed.C.

THE END

OF

ARTISTIC ANATOMY OF

ANIMALS

Zeitfracht Medien GmbH
Ferdinand-Jühlke-Straße 7
99095 Erfurt, Deutschland
produktsicherheit@kolibri360.de